Watchful Lives in the U.S.-Mexico Borderlands

Vigilanzkulturen /
Cultures of Vigilance

Herausgegeben vom / Edited by
Sonderforschungsbereich 1369
Ludwig-Maximilians-Universität München

Editorial Board
Erdmute Alber, Peter Burschel, Thomas Duve, Rivke Jaffe,
Isabel Karremann, Christian Kiening and Nicole Reinhardt

Band / Volume 4

Catherine Whittaker, Eveline Dürr,
Jonathan Alderman, Carolin Luiprecht

Watchful Lives in the U.S.-Mexico Borderlands

—

DE GRUYTER

Funded by the Deutsche Forschungsgemeinschaft (DFG, German Research Foundation) – Project-ID 394775490 – SFB 1369

ISBN 978-3-11-099727-9
e-ISBN (PDF) 978-3-11-098557-3
e-ISBN (EPUB) 978-3-11-098626-6
ISSN 2749-8913
DOI https://doi.org/10.1515/9783110985573

This work is licensed under the Creative Commons Attribution 4.0 International License. For details go to https://creativecommons.org/licenses/by/4.0/.

Creative Commons license terms for re-use do not apply to any content (such as graphs, figures, photos, excerpts, etc.) that is not part of the Open Access publication. These may require obtaining further permission from the rights holder. The obligation to research and clear permission lies solely with the party re-using the material.

Library of Congress Control Number: 2022952250

Bibliographic information published by the Deutsche Nationalbibliothek
The Deutsche Nationalbibliothek lists this publication in the Deutsche Nationalbibliografie; detailed bibliographic data are available on the Internet at http://dnb.dnb.de.

© 2023 the author(s), published by Walter de Gruyter GmbH, Berlin/Boston
The book is published open access at www.degruyter.com.

Cover illustration: Nanzi Muro: *Untitled*, 2022
Printing and binding: CPI books GmbH, Leck

www.degruyter.com

Preface

When Arndt Brendecke, who acts as Speaker of the Collaborative Research Center on "Cultures of Vigilance" at LMU Munich, invited Eveline Dürr to participate in the initial funding application and to contribute to the CRC with a project from anthropology, she instantly thought of her research experiences in New Mexico, where she conducted fieldwork in the late 1990s. Eveline's project in the city of Albuquerque focused on individuals who self-identified mainly as Hispanic.[1] Some took pride in their Spanish ancestry and knew that their families had lived in the area for centuries, while others felt a stronger connection to Mexico and to their Indigenous heritage. However, regardless of their self-identification, most of her interlocutors felt disadvantaged and discriminated against in an Anglo-American-dominated society – and under pressure to justify their presence in their own country. While responses to these conditions of ongoing coloniality differed significantly, questions of belonging and identity were key in their daily lives. In this vein, she became aware of the need to better understand the ways they anticipated being watched and classified as 'other' in an Anglo-dominated context. These observations nurtured her research design for the CRC's project in a setting with different historical trajectories and politics, from which this book on "Watchful Lives" in San Diego, California, results. The objective of the CRC is to discover the ways in which vigilance is employed and has developed historically and across cultures, focusing on how the attentiveness of individuals contributes to collective goals. The aim of our project has been to interrogate the particular ways in which individuals use vigilance in a borderland context in which they face discrimination.

In the original proposal, which Eveline designed before the outbreak of COVID-19, the ethnographic work was center stage and she had planned to carry out joint fieldwork with a postdoctoral researcher. However, in the light of the pandemic, the initial research plan had to be adjusted – and so did the composition of the project team, which eventually evolved to include four researchers. This book then, is the result of the particular circumstances under which it was written. Conducting research at the height of the pandemic further accentuated the very inequalities that we highlight in this book.

The book evolved to have four authors principally because Catherine Whittaker, who was postdoc on the project, took up an assistant professorship at Goethe University in Frankfurt am Main after completing the main phase of the fieldwork in San Diego. We thus also thank Goethe University for contributing research fund-

---

1 Dürr, *Identitäten und Sinnbezüge*; Dürr *¿Héroe Español O Déspota Colonial?*.

ing for Catherine's contributions to this project since April 2021. Catherine's new location and responsibilities required hiring a new postdoc, Jonathan Alderman, who joined at the initial writing-up stage of the project. Meanwhile, Catherine returned to San Diego as often as her teaching schedule and finances allowed, in hope of increasing the accuracy of our descriptions and analyses in a way that serves the communities we are writing about. The fourth author, Carolin Luiprecht, acted as research assistant and then became co-author of this book. She also wrote her master's thesis on a topic related to this research project. Finally, we have been pleased to count on the collaboration of San Diegan artivista (artist-activist) Nanzi Muro, who herself identifies as a *fronteriza* ("borderlander"). Nanzi Muro's non-textual articulations of our book, which accompany each chapter, can be read alongside the texts themselves and as artwork in their own right. We wanted to work with her because she uses art to address social injustice, which is a key topic in this book.

The different members of the team have had different roles in the production of the book. However, while drafts of chapters began with a single author, the book has evolved to a shared authorship over the book as a whole. While this writing process has at times been challenging, it has allowed us to discuss how we each understand vigilance, and what we each took from our experiences in San Diego, as well as allowing us the opportunity to read through and improve each draft of chapters collectively. For these reasons, it is important to introduce the team members individually, our biographies and our roles in the production of the book.

Eveline Dürr is a professor of Social and Cultural Anthropology at LMU Munich, where she is engaged in a range of mostly collaborative research projects on urban issues, such as ethics and notions of the "good life" in cities, "poverty" tourism and inequality, (non)human-environmental entanglements and identity politics. She was trained at Universities in Heidelberg, Mexico City and Freiburg. She has lived and conducted fieldwork in Mexico, the U.S. and in New Zealand. In each case, she pays special attention to political forces and local responses, and the ways their interplay shapes individuals' life worlds. Her roles as Deputy Speaker of the CRC and as Principal Investigator of this project consisted of conceptual work, in particular on vigilance, subjectivation and temporality, in managing the project's workflow and in co-writing this monograph. During our joint fieldwork stay September 2021, Eveline noticed similarities and differences between New Mexico and California pointing to the heterogeneity of the borderlands and the resulting identity politics. In addition, she made suggestions for possible publication avenues for further research outputs and enhanced the project's profile by increasing its international visibility, facilitating workshops and international conferences – all supported generously by the CRC.

Catherine Whittaker is an Assistant Professor at the Institute for Social and Cultural Anthropology at Goethe University Frankfurt, Germany, and was the principal fieldworker in this project as a postdoctoral researcher at LMU Munich (2019–2021). She was drawn to this project both because of her personal life and her research background. Born in Central Germany as the trilingual daughter of Italian and Irish-British-Australian immigrants, Catherine has continued to live a nomadic life. Having often been mistaken for a foreigner in her countries of citizenship sensitized her to issues surrounding migration, mixed identities, and belonging. In her previous research on women's anti-violence activism in Michoacán, Mexico, Catherine had found vigilance to be a salient issue: entering some towns meant being watched, or even stopped, by cartel lookouts and police or armed citizens. Many local people navigated insecurity by ensuring that their neighbors knew them and watched out for them. It follows that where the state is absent or untrustworthy, a culture of vigilance becomes key to safety. After completing a postdoc on the Michoacán research at the University of Aberdeen, UK, in 2019, Catherine was a visiting scholar at the University of California San Diego (2020–2022). She has studied Latin American Studies and Anthropology in Bonn, Oxford, London, Edinburgh, and Mexico City. Currently, she is working on a book on the interconnectedness of love and violence in Central Mexico, based on her 2019 PhD thesis at the University of Edinburgh (2018 recipient of the Radcliffe-Brown/Sutasoma Award of the Royal Anthropological Institute) and a recent article for American Anthropologist, "Beyond the Dead Zone: The Meanings of Loving Violence in Highland Mexico." Her research is driven by the desire to humanize often misunderstood populations, such as survivors and perpetrators of violence, by uncovering the structures that shape their circumstances.

Jonathan Alderman came into the project after Catherine had already conducted a year's fieldwork in San Diego. He has previously carried out ethnographic research in Bolivia. Although new to the project, he had already been interested in some of the themes that became important in the course of the project, such as citizenship, belonging and subjectivity. He studied Philosophy at the University of Essex, completed a master's degree in Latin American Studies at the Institute of Latin American Studies, University of London, and finally a PhD in Social Anthropology at the University of St Andrews in Scotland. His PhD thesis, *The path to ethnogenesis and autonomy: Kallawaya-consciousness in plurinational Bolivia* concerned citizenship and subjectivation within Bolivia as it became a plurinational state. He has conducted postdoctoral research in Bolivia, including as a research fellow at the Institute for Latin American Studies, University of London. This research has mainly examined the relationship between rural Bolivians and their houses and how changes to the materiality of the house impacts the social relationships of its inhabitants. His interest in social housing developed into an interest in infrastructure more widely, through producing an edited book titled *The Social and Political Life of Latin American Infrastructures*. Since he had no first-hand experience of the U.S.-Mexico borderlands, the team fieldwork in September 2021 was useful in broadening his understanding of San Diego and the issues that people face living in the area. This experience enabled him to visualize many of the places that are described in Catherine's fieldnotes and interviews. Coming into the project had its challenges, not only in getting to grips with a new topic, but also a new academic environment (The CRC) and a new place to live (Munich). It has therefore also provided him with great opportunities to appreciate academia from a new perspective and new experiences that come from living in Germany.

Carolin Luiprecht grew up in Northern Italy in a mainly German-speaking region directly bordering Austria. She has thus naturally been interested in borders and border issues for a long time. After finishing school, she went to Munich in Germany to study Social and Cultural Anthropology at LMU, where she completed her master's degree in February 2022. In terms of region, her focus has been on the Americas, especially Colombia and the U.S.-Mexico border. Thematically, she researches migration, activism, and tourism under the rubric of Urban and Border Studies. She also has a special interest in intersectional and post-/decolonial approaches to anthropology. She became part of this project in summer 2019, shortly after completing her bachelor's degree in Social and Cultural Anthropology. Carolin worked for Eveline and Catherine, and later Jonathan as a student and research assistant. This gave her insights into the workings of both a research project and a university. She travelled to San Diego with the whole team in the fall of 2021 to conduct team fieldwork and pursue her own research direction for her master's thesis. For her fieldwork, she carried out digital participant observation on Instagram with a group of self-identifying *brujxs* (witches) and their network. She examined how healing and spirituality relate to social justice activism, particularly concentrating on digital representation of these practices. Interviews with the main group then helped to deepen her understanding of this. During her time in San Diego, Carolin also conducted fieldwork as part of the project team. By assisting within the research project and conducting her own research, Carolin had the opportunity to contribute to this book.

Nanzi Muro has been interested in art ever since she remembers. At just 14 years old, she discovered her passion for photography. So, after finishing high school she naturally began her Bachelor of Fine Arts at the San José State University in San Diego, which she completed in 2019. She currently lives in San Diego where she is studying her master's degree. Since the beginning of her studies, she has been doing professional and independent photography of a variety of subjects. Due to her biographic history (she was born in Los Angeles, raised in Tijuana and lives in the borderland), she identifies herself as *fronteriza*. This special perspective is important for her artistic and activistic work. Thus, her current artistic focus is on culinary photography, bringing out the texture and flavor of food through images and highlighting the variety of typical food and its importance for migrant- or border movements in the borderland. Also, as an artivist she tries to raise awareness of the histories, perspectives, and biographies of the borderland and migrant communities in Mexico and in the U.S. Thus, she organizes and accompanies various social movements and social projects as a photojournalist and an editor. Within her art she advocates for social change and tries to give a voice to the voiceless. She was happy to accept the authors' invitation to provide the art for this book.

This book, then, has been highly collaborative, but not just because of the writing process through which it has evolved. As authors, we would also like to acknowledge the support that we have received, both during the field research and in the writing-up process. We are grateful to numerous people who have helped us to bring this book to fruition. Firstly, we are particularly grateful to the institutions and individuals in San Diego that have oriented us in our understanding of Chicanismo and the U.S.-Mexico borderlands. These include the Anthropology Department of the University of California, San Diego, for acting as host institution during Catherine's main period of fieldwork, and especially Nancy Postero and Rihan Yeh, amongst the faculty of the Anthropology department. We would also like to particularly thank Roberto Hernández and Alex Gomez (both SDSU) for meeting with and sharing their own experiences and understanding with us, Alberto López Pulido for helping us to understand the term *trucha*. We would like to thank Nanzi Muro for enhancing the book by providing her artwork.

In Munich, we are grateful to the Collaborative Research Center on Cultures of Vigilance (CRC 1369) at LMU Munich and the financial support of the German Research Foundation (DFG, Deutsche Forschungsgemeinschaft) for supporting the research project through to its production as a book. At the CRC, we acknowledge the insightful comments received within the working group on subjectivation as well as at the CRC's colloquia and workshops. This has included the workshop that we organized titled "Borderland Vigilance: re-conceptualising borders in comparative perspective" held at LMU Munich in July 2022. We are grateful for the stimulating and thought-provoking discussions on vigilance in borderlands that arose both in the workshop and at panels we have organized at the European Association of Social Anthropologists (EASA), Latin American Studies Association (LASA), and American Anthropological Association (AAA). We would like to thank the panelists of these respective panels for their thoughtful comments, which have furthered our own understanding of vigilance in borderlands. At the CRC we are grateful for the support provided during the publication process, such as copy-editing and indexing. We are also grateful to Daniel Dumas for producing several maps that we have used in the book. We would also like to thank colleagues at the Institute of Social and Cultural Anthropology for insightful comments and ongoing collegial and administrative support, particularly Raúl Acosta for sharing his contacts in San Diego, and Henry Kammler for contributing an analysis of the concept of *trucha* to the CRC blog. We would also like to thank Ana Ivasiuc (Maynooth University) for collaborating with this project, resulting in a special section of the journal *Conflict and Society*.

Finally, this book is the result of a team effort, and, as researchers and writers, we have benefitted considerably from those who have supported and collaborated with us. In particular, we are indebted to Chicanxs who have engaged with our research. With good reason, some people in the Chicanx community can feel uncomfortable with the agendas of researchers they do not know (and even those they do know), for example, Jacob (not his real name), whom we cite in Chapter 2, who was annoyed at what he perceived as researchers building their careers through studying Chicanxs. Others (for example, academics) that the team encountered were interested in and sympathetic to our project. Some people were both interested and skeptical in helping our research apparently because they welcomed inter-racial solidarity. We aimed to be watchful in our writing and were aware of the contradictions in our position, as researchers taking a critical perspective towards structures of coloniality, who ourselves come from an elite university in Europe, and benefit from unequal hierarchies of power. Our privileges as white anthropologists are founded on coloniality, even as we seek to visibilize and destabilize racism and the structures that underpin and perpetuate it, and the ongoing struggles to chip away at the coloniality that Chicanxs and other People of Color face on a daily basis. We have attempted to hold these contradictions in tension, but such contradictions are not easily resolvable – if at all. However, reflecting on our own positionality, and the watchfulness of people in Barrio Logan and elsewhere in San Diego towards us as researchers, has helped us to consider more deeply the watchfulness of Chicanxs and other racialized and migrantized people in the borderlands. We hope that this research may contribute to their decolonization struggles and highlight pathways to healing. We also hope that this work contributes to discussions of vigilance in borderlands, as well as an understanding of the continuing struggles of Chicanxs and other racialized people in the U.S. for equal rights as citizens.

The authors
Munich, November 2022

Figure 1: From left to right, Catherine Whittaker, Jonathan Alderman (back), Eveline Dürr (front), Carolin Luiprecht.

Contents

Chapter 1
Introducing Watchfulness in San Diego —— 1
    Watchful subjects —— 6
    Understanding of the self and anticolonial subject formation —— 13
    A borderland city —— 20
    Coloniality in San Diego —— 29
    Structure of the book —— 32

Chapter 2
A Hall of Mirrors: Watchfulness as Ethnographic Method —— 35
    Suspicious eyes meet Anthropology's colonial gaze —— 38
    Learning whom to ask —— 41
    The Border as mirror multiplied —— 45
    Room hunting in a landscape of anxiety —— 48
    The upheavals of COVID-19 —— 50
    Relationship rupture and repair —— 53
    Positionality, intersectionality, and emerging research subjectivities —— 57
    Group fieldwork —— 59
    The writing process —— 62

Chapter 3
"Sometimes you have to Transform into a Serpent": Political Subject-making around Chicano Park —— 67
    Becoming Chicanx subjects under the Coronado Bridge —— 70
    Citizenship-in-action: Building Chicano Park and creating community —— 73
    The aesthetic politics of Chicano Park's murals —— 78
    Imminent danger from the bridge —— 84
    "Ponte trucha" —— 86

Chapter 4
Watching out across Time and Space in Aztlán: Chronopolitics in Chicano Park —— 91
    Watching and being watched as political subject formation —— 92
    Chicanismo as an intergenerational project: intercommunity linkages and ruptures —— 98

Chicano Park as Aztlán: Challenging colonial understandings of time and space —— 101
Performing the struggle across time and space —— 107

Chapter 5
"Why us?": Making Environmental and Health Threats Visible —— 115
An autoethnographic perspective on the COVID-19 pandemic in San Diego —— 120
The intersectionality of vulnerability and responses to the pandemic —— 124
The pandemic's impact on struggles against gentrification and eco-colonialism in Barrio Logan —— 130
The colonialist militarized apartheid logic of environmental health —— 135
Watchfulness in extraordinary conditions —— 141

Chapter 6
Watchful *Brujxs:* Social Justice Activism in the Digital Sphere —— 147
Healing and social justice —— 150
Brujxs and watchfulness on social media —— 155
Shadow work —— 158
Calling out —— 161
Ongoing watchfulness and healing for social justice —— 164

Conclusion
Decolonial Watchfulness and Watchful Lives in San Diego —— 167
Trucha! —— 169
From vigilance to vigiculture —— 174
Toward a watchful anthropology —— 177

References —— 179

List of Acronyms —— 196

List of Figures and Illustrations —— 197
List of Figures —— 197
List of Illustrations —— 198
List of Maps —— 198
Lists of Charts —— 198

Chapter 1
Introducing Watchfulness in San Diego

"I was wondering if you're an FBI agent," Pepe[1] admitted to Catherine some weeks after they first met at a vigil in March 2022. The vigil was commemorating the life of a highly respected and well-loved elder from the San Diegan Brown Berets, who had also been one of the founders of this semi-militant community defense organization. Since the national organization's beginnings in 1966, the Brown Berets had frequently been under FBI observation,[2] and Pepe had encountered a few himself. Thus, as a white European stranger asking many questions, Catherine had seemed suspicious to him. The twenty-something-year-old U.S.-born son of Mexican parents, a *Chicano*, maintained a serious expression throughout their chat. Yet despite his skepticism, Pepe was willing to connect with the 33-year-old researcher from Germany, telling Catherine about his plans to start a gun club and join the Marines, while fittingly donning a black tactical t-shirt, dark cargo pants, and a brown beret on his black curls. As a Beret, he would often act as security at community events. Yet even before joining the organization at the age of 17, he had learned to be watchful growing up in a neighborhood rife with "violent masculinity" and gangs near several detention facilities and the world's most militarized border. While cruising in his neighborhood, he pointed to a McDonald's restaurant in San Ysidro that had been the site of a grisly massacre in 1984. A white Anglo-American man dressed in a military-style outfit, apparently frustrated by U.S. losses in Vietnam and blaming Mexicans for his unemployment, killed 21 mostly Mexican and Mexican American people with a semi-automatic rifle until he was finally shot by a sniper. According to Chicanx Studies scholar Roberto Hernández, who lived two blocks from the McDonald's at the time of the massacre:

> The trauma of the event was inscribed [...] collectively on San Ysidro residents who [...] recognized the colonial/racial dynamic that informed the shooting. In this sense, it was arguably reminiscent of the systematic killings of Native peoples that led to the eventual creation of the border and the current globalizing phase of modernity/coloniality, as many Mexican@s knew all too well that numerous frontier massacres had seen more than twenty-one people killed at once.[3]

---

1 All interlocutors' names in this book are pseudonyms.
2 Correa, The Targeting.
3 Hernández, *Coloniality of the US/Mexico Border*, p. 115.

Since then, a slew of white supremacist mass shootings, including the El Paso shooting in August 2019, in which 23 mostly Mexican and Mexican American people died,[4] as well as the rise of anti-migrant vigilante groups, ongoing police violence against working-class People of Color and militarized reactions to Black Lives Matter protests[5] have contributed to a sense of the U.S.-Mexico border as "a war zone."[6] That is not to mention the violence of the highly-surveilled and patrolled border wall itself, which cuts Pepe off from "the other half of my city," as he conceives of San Diego and Tijuana as one culturally and economically interlinked unit.

Beyond these spectacular displays of violence and threat,[7] less visible forms of state violence and neglect have been no less damaging to Pepe's community. Himself currently an undergraduate student, Pepe recalled that before 2002, San Ysidro did not even have its own high school, so that many students did not graduate and had limited prospects, accordingly, often having to work under exploitative conditions in physically taxing and underpaid jobs. Even after twenty years living north of the border, Pepe's mother struggled financially. Shortly after her third child, Pepe, was born, she was deported as an unauthorized migrant. Years later, she discovered that she had been sterilized in the detention center without her consent – a eugenicist practice that continued in some Californian prisons and detention centers until the prison anti-sterilization bill[8] was signed into law in September 2014.[9]

"We were never meant to survive," Audre Lorde wrote to encourage other Women of Color to speak up against injustice.[10] Who do you become, when the system seeks to surveil, harass, incarcerate you, when it even seeks to prevent you from being born? In the face of such a politics of death, a necropolitics,[11] how do you respond?

Pepe responded by working towards revolution. His preparation for it and everyday anticolonial resistance consisted in training, educating, and arming himself, while looking out for his community. For him, watchfulness was a way of life.

In this book, we show how watched and 'othered' people in the border city of San Diego on the Southwestern edge of the Unites States respond to racism and

---

4 Aguilera, Mass Shooting, El Paso.
5 Whittaker/Dürr, Vigilance, Knowledge, and De/Colonization.
6 Hernández, *Coloniality of the US/Mexico Border*, p. 109.
7 Cf. Valencia Triana, *Capitalismo Gore*.
8 SB 1135. http://www.leginfo.ca.gov/pub/13-14/bill/sen/sb_1101-1150/sb_1135_cfa_20140620_180402_asm_comm.html [last access: 10/07/2022].
9 Ray, California's central role in the eugenics movement.
10 Lorde, *Your Silence*.
11 Mbembe, Necropolitics.

surveillance with varied practices of watchfulness. As we will show, watchfulness goes beyond counter-surveillance, as it constitutes an integral part of many people's daily lives and shapes their individual and collective subjectivities within contemporary U.S. society. Experiencing racist discrimination often leads to developing a vigilant disposition, which in turn becomes a significant aspect of people's everyday practices and what it means to be living in the borderland around the physically divided twin cities of San Diego and Tijuana. Focusing particularly on Chicanxs (Chicanos of all genders), but also including other disadvantaged and racialized individuals and groups, we explore how individuals intervene against structural inequalities and threats in their lives, such as by re-claiming space, consciousness raising, participating in protests, and engaging in healing practices. As we will show ethnographically, to be Chicanx and San Diegan is intimately connected to the situation of living in a borderland condition defined by inequality and violence, an in-betweenness created by the meeting and mixing of different cultures that is more than the sum of its parts.[12] The borderland character of San Diego emerged through the incorporation of California into the United States as an outcome of the Mexican-American War of 1846–1848. Our ethnography looks at how present-day U.S. citizens, who are Chicanx, Latinx and otherwise watched and 'othered,' deal with being treated with suspicion and as aliens on their own land. We argue that contestations surrounding belonging create particularly watchful selves and that this is more broadly a significant aspect of borderland lifeworlds. Watchfulness is an ambivalent practice that can express anxiety, but also care and contribute to community building. As we will show in this ethnography, watchfulness can represent a way of life.

Despite its image of a relaxed vacationland, San Diego is also a highly militarized, conservative border city. As most unauthorized immigration to California took place in the San Diego area in the 1980s, the city responded with different forms of border security and has continued militarizing its border ever since.[13] According to the memoir of Francisco Cantú, a former border patrol agent, the militarization of the border dehumanizes both those trying to cross and those defending the border.[14] Although San Diego is also a "sanctuary city," which means that local police have instructions not to provide information to Immigration and Customs Enforcement (ICE), and California is a multi-ethnic state famous for its progressive politics, in practice, ICE terrorizes local mixed-status neighborhoods. For instance, Berenice, a student in her late twenties who was born and raised in San

---

12 Anzaldúa, *Borderlands/La Frontera*.
13 Davis/Mayhew/Miller, *Under the Perfect Sun*.
14 Dürr/Whittaker, Introduction, p. 5.

Juan, Puerto Rico, grew up having to be attentive because she had to look after her mother, who lives with epilepsy and diabetes. Berenice described sometimes anxiously checking whether her mother was still breathing at night. More recently, she persuaded her mother to move in with her in San Diego, in order to receive better health care than was available to her on the island – a fact which Berenice argues is rooted in structural racism. When her mother was hospitalized the summer of 2019:

> I was just like, why is [this ICE SUV] there – I thought it was just circumstantial, like someone who had just been detained was just taken to the hospital. And then I realized it was in front of the ER all day every day, multiple days in a row. [...] I cannot fathom, like, me taking my mom to an emergency in a life-or-death diabetic crisis and not having documents and then there's an ICE thing there and then what do I risk? Being separated from my mom, her dying [...] alone without help and me being taken to a freaking detention center?

After picturing this alternative future for her family and the real threat to unauthorized immigrants in her community, Berenice photographed the vehicle and uploaded it to a community advocacy organization's social media page in order to warn others. In this way, Berenice directed her vigilance towards anticipating potential harms to her wider community. By examining this kind of peer-to-peer vigilance beyond top-down surveillance and bottom-up sousveillance,[15] we go beyond previous conceptualizations in Security Studies. Thus, we argue against a unidirectional understanding of vigilance as vertical, horizontal, or lateral, but rather stress the crossovers, crisscross connections and relationality of watchfulness. Our book therefore aims to rethink watchfulness as a multilayered concept, always entangled with its socio-cultural context more broadly. Further, we show the consequences of internalized watchfulness and discuss the practices that result from these processes. In doing so, we advance the theoretical discussion on watchfulness in anthropology, while also refining and expanding the conceptual toolkit of Security Studies.

In the following, we introduce some of the key concepts that we will be using in our ethnographic analysis of watchful responses to contemporary inequalities and discrimination in San Diego. Throughout the book, we pay particular attention to notions of vigilance in the specific context of borderlands and coloniality, and we are interested in understandings of the self- and subjectmaking in relation to temporality and racialized, othered bodies.[16] We also stress the Chicanx "struggle" fighting against injustices and coloniality, its dynamics over time and extension

---

15 Wolverton, "Sousveillance."
16 See also Dürr et al., *Becoming Vigilant Subjects*.

into the digital sphere. Further, we draw on the historical context through which San Diego-Tijuana developed as a borderland and what kinds of subjectivities emerged among racialized people on the U.S. side of the border. Citizens and lawful residents like Pepe, who identifies as "Mexicano," "Chicano," "Indigenous," and "Cholo" (counter-cultural Chicano youth with Indigenous ancestry), often find themselves stereotyped by white Anglo Americans as well as racially profiled and criminalized by law enforcement. This is almost inevitable as already in their training, police recruits learn to perceive and engage with racialized and gendered bodies according to specific "scripts" that ultimately enable racialized police violence.[17] Those whose appearance does not always differentiate them from newly arrived migrants often face disadvantages – even if they have lived in the United States for many generations. This includes invisible threats, such as from environmental pollution and unequal health care provision, which we discuss in chapter 5. As we argue throughout the book, being, in Anzaldúa's terms, borderlanders (*fronterizxs*) and crossers of the border that crossed them (*transfronterizxs*),[18] and the everyday struggles this entails, involves a level of personal and collective vigilance that marks the subjectivity of Chicanxs and members of other similarly racialized and disadvantaged groups. Throughout this book, we extend our analysis to include other non-white, colonized people who experience discrimination in the U.S.-Mexico borderlands and San Diego more specifically. Many of the Chicanxs we worked with identified as *Raza*, which they understood as an inclusive term to describe Latinx, Indigenous, and other People of Color. However, some people argue that the term erases Black people and excludes Indigenous people because it refers to José Vasconcelos' concept of a *"Raza Cósmica,"*[19] which is thought to be created through the mixing of Spanish and Indigenous people in Mexico, producing a superior mestizo subject. This is particularly important for chapter 6, where we ethnographically explore the digital watchfulness of healers and self-identified *brujxs*[20] who do not frame themselves as Raza but rather criticize and reject the term, and where we examine their healing practices addressing the manifold consequences of exclusion. Resistance against coloniality and white

---

17 Aushana, Inescapable.
18 See Fránquiz/Ortiz, Who are the transfronterizos.
19 Vasconcelos, *Raza Cósmica*.
20 *Brujxs*, from *bruja/brujo* (witch/witcher in Spanish) is a non-gendered self-designation. While it is commonly used in a derogatory and accusatory way, activist Latinx and Chicanx groups and especially women and non-binary people try to reclaim it.

supremacy, as well as other experiences of discrimination plays a pivotal role in everyday life for Black and Indigenous activists, as well as other activists of Color.[21]

Watchful subjects

"If you see something, say something": This famous public awareness campaign encapsulates what vigilance is typically about. It was originally introduced in New York City after the September 11 terrorist attacks.[22] Depicted here is a recent bilingual English and Spanish version as seen on a San Diegan trolley in February 2020 (see Fig. 2). Such slogans encourage citizens to support and collaborate with law enforcement by observing their surroundings attentively and reporting anything suspicious. In this way, citizens are made responsible for their collective safety, rather than just leaving safety in the hands of the police. This also means that the state uses its citizens to comply with tasks of the state, such as providing security, and thus turns watchfulness into a form of governance. Appeals to be watchful are made meaningful by tying them to values that point beyond self-interest, portraying them as being of great social concern. However, the state does not always succeed in channeling the effects of observations in the desired direc-

Figure 2: Public Service Announcement on a San Diegan trolley, February 2020.

---

21 We capitalize Raza, Black, Brown, Indigenous and People of Color, following a wider consensus to do so in U.S. scholarship. This spelling highlights the constructedness of these protected categories, thus helping to denaturalize pseudo-biological and discriminatory assumptions about what are, stereotypically and without scientific basis, often considered to be different "races" in popular thought. White, however, is not capitalized in this book, as people racialized as white do not share the same experiences of racism as people who are racialized differently.
22 Fernandez, A Phrase for Safety. See also Emerson, Vigilant subjects.

tion – rather, individuals may choose actions which can challenge or even oppose state interests.²³

As Arndt Brendecke and Paola Molino suggest, many fundamental social services like security rely "on observations made and communicated by regular citizens who neither observe 'from above' nor are representatives of any particular institution."²⁴ Vigilance thus consists in "services rendered by people who willingly report what they have seen, heard or sometimes smelt,"²⁵ linking individual attention with institutional tasks, such as security. The authors argue that interaction between private attention and broader institutions is mediated by complex cultural, linguistic, and social relations, which shape the ambivalent civic self, a process of subjectivation. According to Jacques Rancière, civil society is thus key to the functioning of "the police," which he conceives not simply as an institution of the state, but as a widespread disposition to maintain a particular social order. In Rancière's sense, the police includes ordinary Anglo-American citizens, some of whom patrol the border area as vigilantes.²⁶ On the other hand, marginalized U.S. citizens, including working-class People of Color, are often the implicit object of vigilance. As analyses of the "If you see something, say something" campaign have shown, what people perceive as suspicious often is directed by their conscious and subconscious biases, thus drawing on racialized imaginaries of threat.²⁷ Thus, seeing, which is the privileged sense in the context of vigilance, is not a neutral, universal human sense, but profoundly shaped by our socialization.²⁸

Vigilance is a concept deeply associated with racism and vigilantism in the U.S., which in turn inspires watchfulness on behalf of those watched. At the U.S.-Mexico border, patrols of vigilantes known as Minutemen search for anyone who has crossed the border illegally.²⁹ Vigilantism refers to "taking the law into your own hands without any legal authority."³⁰ However, it is not just an increase in vigilance, but a premeditated form of action by private citizens who use force or threat to (as they see it) provide assurances of security which they feel are otherwise lacking.³¹ Dubbed the "vigilante president" by *The New Republic*, former pres-

---

23 Dürr, Beobachter:in.
24 Brendecke/Molino, The Cultures of Vigilance, p. 11.
25 Ibid.
26 See also Hernández, *Coloniality of the US/Mexico Border*.
27 Balme, Hypervigilance.
28 Ivasiuc/Dürr/Whittaker, The Power and Productivity.
29 Walsh, Community, surveillance and border control; Shapira, *Waiting for Jose*; Arfsten, *The Minuteman*, Auf der Jagd.
30 Mareš/Bjørgo, Introduction, p. 1.
31 Johnston, What is Vigilantism?

ident Donald Trump infamously defended right-wing vigilante groups,[32] while warning against immigrants, drugs, crime, and COVID-19, which he called the "China Virus," crossing the southern U.S. border.[33] In response, Trump increased numbers of border patrol agents and ICE raids, while upgrading technological border surveillance measures. The high number of border patrol agents (over 21,000 in total at the time of writing this) means that Mexican Americans living near the border regularly see people looking like themselves being chased by border patrol agents.[34] Unsurprisingly then, many San Diegan working-class Latinxs associate the word "vigilance" with racist vigilantes and hostility towards themselves and (other) People of Color.

Vigilance and its related concepts, watchfulness and attentiveness, and subjectivation are central to our research. To be vigilant describes individually directing condensed attention towards an externally set target either to navigate a situation of heightened uncertainty or to avert a specific perceived danger in the service of an assumed greater good, which may include social, moral, or religious goals.[35] In the case of Chicanxs and other groups mistaken for migrants, the perceived danger may present itself in the form of *la migra*, the immigration agencies of the U.S., as there have been cases in which U.S. citizens have been unlawfully deported for not carrying proof of residence when encountering border patrol officers.[36] In this book, we examine how people living in a cultural and political borderland, particularly people who are not themselves migrants, develop practices of vigilance that help them to navigate daily life. Furthermore, we are interested in the concomitant processes of subjectivation that result, as people anticipate racial discrimination and violence against themselves in order to avoid or minimize it.

Unlike surveillance, vigilance has yet to be examined more closely by anthropologists. The concept has often been used vaguely[37] and been employed interchangeably with alertness, caution, anxiety, fear, and tension.[38] In one of the few articles directly engaging the concept, Henrik Vigh argued that particular environments of uncertainty, where an enemy may not be easily identifiable, produce "a constant awareness and preparedness toward the negative potentialities of social

---

32 Hurst, The Vigilante President.
33 Hee Lee, Donald Trump's false comments.
34 Hernández, *Coloniality of the US/Mexico border*.
35 Brendecke, Attention and Vigilance, p. 17.
36 Hernández, *Coloniality of the US/Mexico border*, p. 115.
37 Wolf-Meyer, Editorial Introduction: Alertness.
38 Crane/Pascoe, Becoming Institutionalized; Regnier, Forever slaves?, McKenna, "We're Supposed to Be Asleep?"

figures and forces."³⁹ Contrasting the ethnographic contexts of Bissau and Belfast, he observed people constantly employing all their senses to scan their surroundings "for early warning signals, creating a heightened awareness toward the unfolding of social and political life."⁴⁰ This heightened awareness can be oversensitive, meaning that in a context of ongoing conflict any perceptible sign of difference, even as mundane as a haircut, can come to be perceived as a threat. In passing, Vigh acknowledges that vigilance represents a "struggle to gain clarity and knowledge of these invisible yet dangerously present" threats.⁴¹ In our work, we draw on the "negative potentialities" highlighted in Vigh's study that may create a hypervigilant self, but we also take it further by showing the flip side to this, by carving out the enabling and empowering potential of watchfulness. This is a characteristic particularly of feminist vigilance, which Sotirin describes as going beyond defensive aspects to combine anger with hope. Thus, feminist vigilance, "fueled by anger and critique"⁴² emerges from an "affective landscape dominated by anxiety, fear and suspicion"⁴³ to create new ways of thinking, of being, and possibilities for change.

Watchfulness, as one of the terms we use to describe vigilance by racialized people throughout this book means watching over others as well as oneself. It is not only a "way of seeing," but also "a way of being."⁴⁴ In this sense, Daniel Goldstein has defined it as "an alert disposition through which barrio residents hope to protect themselves from harm by spotting it before it strikes."⁴⁵ Based on his research in Andean Bolivia, Goldstein argues that such watchfulness is not only reactive but is also "generated by insecurity and a sense of abandonment – of being left outside the law's protections, or out-lawed."⁴⁶ Similarly, we are interested in the watchfulness of people who are non-state actors, particularly over their own behavior,⁴⁷ in contrast to the surveillance of police or officials representing organs of the state. Rather, we are interested in individuals who have become aware of being under surveillance and practice watchfulness in response. We therefore ask: How do the watched become watchful?

---

39 Vigh, Vigilance, p. 99.
40 Ibid., p. 104.
41 Ibid., p. 110.
42 Sotirin, Introduction to Feminist Vigilance, p. 12.
43 Ibid., p. 10.
44 Finn, Seeing Surveillantly.
45 Goldstein, *Outlawed*, p. 122.
46 Ibid.
47 Frekko/Leinaweaver/Marre, How (not) to talk about adoption.

Colonizers have historically observed and inspected the colonized in order to control them. As surveillance conditions shape their everyday actions, the colonized become watchful in turn – not only towards state representatives, who may act friendly or hostile in different situations, but also of their own behavior. Monitoring themselves is an expression of having embodied the "colonial gaze," a concept popularized by postcolonial authors, such as Frantz Fanon (1970) and Edward Said (1978). According to Fanon, Black people in colonial and postcolonial Algeria came to embody the "white gaze" of the colonizer, as it physically conditioned the ways in which they moved in their environment.[48] Fanon describes this as a process through which attempts to meet colonizers' expectations alienate people from their own self, which is eventually colonized. In a similar vein, W.E.B. DuBois argued that Black people in early twentieth century U.S. formed a "double consciousness," as they had learned to anticipate how their appearance and conduct might be perceived by any white people present.[49] However, as we argue in chapter 4, these processes work somewhat differently and produce different consequences in the Chicanx context. Chicanxs see their ancestral homeland as overlapping with U.S. territory, and thus see themselves as colonized people while also acknowledging their own involvement in colonial practices – what Chicana feminist scholar and poet, Gloria Anzaldúa has framed as "new consciousness."[50]

Nevertheless, these examples demonstrate that over time, colonized people typically come to embody, both in the sense of enacting and subconsciously carrying with them, a sense of being watched by colonizers or other institutional and social forces. This embodiment of the colonizer's gaze is fostered by colonial institutions and partially replaces state surveillance. Anzaldúa refers to this embodied alertness as *la facultad*: "[...] we are forced to develop this faculty so that we'll know when the next person is going to slap us or lock us away. [...] It's a kind of survival tactic that people, caught between worlds, unknowingly cultivate. It is latent in all of us."[51] Cultivating *la facultad* allows people to instantly, intuitively sense "the deep structure below the surface" on a subconscious, pre-verbal level, thereby losing their ignorance and innocence,[52] which finally leads to a permanent state of vigilance.[53] Accordingly, *la facultad* is a multi-sensory, embodied survival tactic that goes beyond mere awareness and preparedness, as it enables Chicanxs to recognize coloniality and resist it, such as by anticipating discrimina-

---

48 Nielsen, *Foucault, Douglass, Fanon, and Scotus.*
49 DuBois, *The Souls of Black Folk.*
50 Anzaldúa, *Borderlands/La Frontera*, pp. 77–98.
51 Ibid., p. 39.
52 Ibid., p. 38.
53 Anzaldúa, *Borderlands/La Frontera*, p. 51.

tion or refusing to assimilate into Anglo-American society. Nevertheless, the privileging of *mestizaje* in Anzaldúa's work has been criticized for erasing Black and Indigenous survivance[54]. Anzaldúa drew heavily on the work of Mexican nineteenth century writer José Vasconcelos, whose description of *mestizaje* in Mexico as creating a "cosmic race" privileged whiteness within the mixture of races and assumed the disappearance of native peoples and then placed Black people at the bottom of a cosmic hierarchy.[55] In addition, Anzaldúa has been criticized for not fully acknowledging certain power hierarchies by claiming that through the new mestiza consciousness individuals can resume power by transcending them.[56]

As we show in our work, vigilance is also central to other disadvantaged actors' self-understanding because their multiple cultural affiliations as well as their ambivalent relationship with the past demand a comprehensive, critical awareness of history as well as an intensive examination of their own positionality. This intense reflexivity plays out in particular watchful practices towards the self – for instance, when racialized people engage in shadow work as we show in chapter 6. Here, we describe how traumatic experiences as well as stereotypes are addressed in order to cope with the consequences of coloniality – a process which the actors frame as "healing." The constant negotiation of belonging, living at cultural crossroads, and the never-ending identity work involved can become a continuous mental and psychological challenge in everyday life for some people in the U.S.-Mexico borderland. On the one hand, as Anzaldúa describes, this can trigger stress, worry, and pain, but it can also act as a breeding ground for agency and a capacity to navigate tense situations competently in different cultural contexts.[57]

This multifaceted watchfulness is referred to as "being *trucha*"[58] by our Chicanx interlocutors and framed as "a way of life." We argue that "to be *trucha*" is a significant characteristic of the sense of self of Chicanxs and the condition it describes can similarly be applied to other racialized people living in the borderland. Accordingly, we argue that this kind of watchfulness is key in their subjectivation processes. It describes not just a negative disposition of vigilance, in the sense of avoiding danger, but also vigilance in the sense of taking care not to harm others in one's vicinity. Through ethnography we show the significance of watchfulness as a practice in a process of racialized non-white subjectivation. The concept of subjectivation highlights the process of becoming a subject. Subject formation can be-

---

54 Vizenor, *Manifest Manners*.
55 See Palacios, Multicultural Vasconcelos.
56 Cuevas, *Post-Borderlandia*, p. 11.
57 Cf. Hammad, Border Identity Politics.
58 See Kammler, Trucha, discussing the meaning of this expression from an ethno-linguistic point of view.

come evident through an act of speech by specific actors, or a practice that takes place in a specific location or under specific material conditions. Elenes argues "that it is precisely when Chicanas and Chicanos became speaking subjects who are politically engaged, naming their own realities, and offering truly democratic alternatives that we [Chicanxs] become dangerous."[59] Challenging who has the right to speak and acting as if one already has the rights that have been denied to them is for Rancière one of the fundamental aspects of what he understands by politics.[60] We argue through the book (particularly in chapter 3), that actions on a collective level to assert their rights to make decisions over their own community and neighborhood have been significant in the formation of Chicanxs as political actors.

It follows that vigilance can have both colonizing and decolonizing effects,[61] as it often arises in response to surveillance and vigilantism: the watched become watchful. As a particularly tightly surveilled city, with "smart lights,"[62] drones,[63] ring doorbells,[64] and neighborhood watch apps[65] producing data for the police,[66] San Diegan infrastructure projects a particular specter of a powerful, all but omniscient state "as imagined, envisaged, anticipated, and ultimately embodied by migrants."[67] Old and new surveillance technologies are key features of borderlands more broadly,[68] collecting footage that is then analyzed with face recognition software. Both the software itself and the databases it draws from have been shown to exhibit racist biases.[69] For Hernández everyday surveillance and harassment through border agents and new technologies are aspects of a larger structure that he refers to as the Civilization of Death.[70] It is an assemblage of oppressive institutions and systems, including Capitalism, Christian religion, Heteropatriarchy, and White Supremacy. Hernández argues that there is a "decolonial imperative" to resist and dismantle this oppressive structure.[71] We are interested in how the spec-

---

59 Elenes, Border/Transformative Pedagogies, pp. 258 f.
60 Rancière, *Dissensus*, pp. 36 f.
61 See also Dürr/Whittaker, Wachsamkeit als Alltagspraxis.
62 Marx, San Diego Smart Streetlights.
63 Zevely, Why a drone may save your life.
64 Fung, Amazon's Ring.
65 Makena, Inside Nextdoor's 'Karen Problem.'
66 Abril, Drones, robots, license plate readers.
67 Barenboim, The specter of surveillance, p. 80.
68 Stop LAPD Spying! 2021; Domingo Garcia, I'm a Mother of Four; Couldry/Mejias, Data Colonialism; Browne, *Dark Matters*.
69 Raji, Data encodes systematic racism.
70 Cf. Hernández, *Coloniality of the US/Mexico Border*.
71 Ibid.

ter of this Civilization of Death may be embodied as a state of constant watchfulness by those who are not necessarily themselves unauthorized migrants but are nevertheless treated by representatives of the state as being under suspicion of being so.

Watchfulness against coloniality is key in creating a cognitive and embodied knowledge, or *conocimientos*, in Anzaldúa's terms, to combat these hegemonic powers.[72] For example, knowledge that police violence disproportionately affects non-white people makes many Latinxs and disadvantaged individuals vigilant when interacting with the state and its institutions, such as the police or migration agencies (see chapter 5 and 6). This same watchfulness is on the one hand a product of coloniality, and on the other contains the potential for decolonization, in that it functions as a warning signal and prevents assaults through attentive observation.

In order to understand the impact of historical conditions on vigilance in the borderlands and for Chicanxs and Latinxs as well as other racialized Black, Indigenous, and People of Color, it is necessary to understand their community formation and self-understanding. For instance, Chicanx self-understanding is characterized by heterogeneity, *mestizaje* and resistance to colonialism and its temporalities. Specifically, we show in this book the ways Chicanx emerges as a community by challenging their reality, and the structures of coloniality that had defined it, as well as extending this understanding to other anticolonial subjects.

Understanding of the self and anticolonial subject formation

To acknowledge and uplift these different perspectives and to pay special attention to the multiplicity of experiences, we apply an intersectional lens to our research and throughout this book. As Hill Collins writes, the concept of intersectionality makes intersecting systems of power and their interconnection with equally overlapping social inequalities visible.[73] While intersectionality is criticized for simplifying subjective experiences and differences and for its essentialism,[74] it is credited for bringing along analytical sensitivity for recognizing sameness and difference in relation to power.[75]

---

72 Anzaldúa, *Borderlands/La Frontera*.
73 Hill Collins, *Intersectionality*, p. 43.
74 Anthias, Translocational Positionality.
75 Cho/Crenshaw/McCall, Toward a Field of Intersectionality Studies, p. 795.

The people at the center of our research share certain experiences of discrimination – being racialized and subjected to racism, (wrongly) being identified as migrants, due to their legal status, or just in general not being read as belonging to a white, heteronormative society. They would mostly identify as non-white as resistance to the persistence of coloniality. Adding to this, the Latinx community is often the target of anti-migratory rhetoric and the criminalization of migration, which can lead to fear and distrust of public institutions.[76] These experiences lead to differing responses: While some might try to blend into society, others resist discrimination in their everyday lives or as activists. Hill Collins credits individuals particularly affected by the oppression of racism, heteropatriarchy, and colonialism with attempting to make power structures visible and developing alternative critical social theories and resistance projects.[77]

Throughout this book, decolonization plays an equally important role: we and our interlocutors do not use it simply as a buzzword,[78] but to resist the continuation of coloniality,[79] thus aiming at radically reconstructing knowledge, power, being, and life itself: "'decoloniality,' understood as the simultaneous and continuous processes of transformation and creation, the construction of radically distinct social imaginaries, conditions, and relations of power, knowledge."[80] The actors we present in the course of this book are either racialized and face discrimination and stereotyping because of phenotype, or they self-identify as Chicanx and/or Latinx, as Black, Indigenous or other People of Color, or more broadly as Raza, in some way.

The Chicanx subject has emerged out of, and is intimately connected to, the condition of the borderlands, but definitions vary. According to San Diegan Chicanx Studies professor, Alberto Pulido, Chicanx "describes Mexican-origin people from the late 1960s through the 1980s who were activists for civil rights and social justice. Chicanismo is the philosophy to commit oneself to live and uphold the values and vision of the Chicano movement."[81] He distinguishes the term from Mexican Americans and Mexicans, while explaining that these terms are often used interchangeably in the borderlands "and speak to the raíces, or 'roots' of this ethnic

---

76 Chavez-Dueñas et al., Healing Ethno-Racial Trauma, p. 54.
77 Hill Collins, *Intersectionality*, p. 117.
78 Tuck/Yang, Decolonization is not a metaphor.
79 Quijano, Coloniality of Power.
80 Walsh, "Other" Knowledges, p. 11.
81 Pulido/Reyes, *San Diego Lowrider*, p. 3. When referring to the political movement, we follow the common usage in the literature of writing "Chicano movement," while using "Chicanx" when referring to present-day people and culture.

Figure 3: Barrio Logan gateway sign.

community."[82] By contrast, for another San Diegan Chicanx Studies professor, Roberto Hernández, Chicanx refers not to "an ethnic identity but rather [...] to politically self-identified individuals or collectives, in keeping with the politics of self-naming that guided its usage in the Chican@ Movement period."[83] Similar to this latter definition, the Brown Berets de Aztlán of San Diego – who might be described as the Chicanx equivalent of the Black Panthers organization – named their monthly self-published print release "Stay Brown" as a "self-reminder to stay true to ourselves, to our community & to our indigenous ancestors, to decolonize ourselves," as they stated in the first issue.[84] In the process, they defined "Chicano" as a non-racially determined "mindset that we use personally to uplift ourselves and we want to share that mindset with anyone who wants to understand & connect with us."[85] For instance, the Barrio Logan gateway sign (Fig. 3) is meant to "pay homage to Kumeyaay, Aztec, Mayan, and all other cultures" and a mural at the Barrio Logan trolley station displays a revolutionary slogan from the 1970s which is widely used in Latin America, "El pueblo unido jamás será vencido" (a united people will never be defeated, Fig. 4). Similarly, for the Brown Berets,

---

82 Ibid.
83 Hernández, *Coloniality of the US/Mexico Border*, p. 33.
84 Brown Beret National Organization, *Stay Brown*, page number unknown. (The photograph of the cited page was shared on the BBNO's Facebook page.)
85 Ibid.

Figure 4: Barrio Logan trolley station mural: *El pueblo unido jamás será vencido* (A united people will never be defeated).

and many others, Chicanx is a non-essentialist identity that is grounded in a history of struggle and obligation to one's community—in never forgetting where one comes from. The word Chicano has also historically been used by wealthier Mexican Americans towards Mexican Americans with indigenous roots; however, particularly since the 1960s, Chicano/a or Chicanx has been used as a self-identifier with positive connotations associated with the Chicano political movement.[86]

Chicanx studies scholars have described the Chicanx subject as forged through the struggles of urban and farmworker unions that from the beginning of the twentieth century fought for improvements in the material basis of their daily work and living conditions and did so in solidarity with other migrant workers.[87] Scholars of Chicanismo (the Chicano movement) have also highlighted the importance of interwoven, cross-border struggles, which have made them "multiply insurgent."[88] These included feminist Chicana movements such as *las Chicanas de Aztlán/Hijas de Cuauhtémoc*, formed at California State University in 1968.[89] Blackwell describes these Chicanas as defining their role as *mujeres de lucha* (women in struggle).[90] We argue that the struggles that we describe in the book, particularly

---

[86] Oliver, Race Names.
[87] Acuña, *Occupied America*.
[88] Blackwell, *¡Chicana Power!*, p. 27; Acuña, *Occupied America*.
[89] Blackwell, *¡Chicana Power!*.
[90] Ibid., p. 61.

the actions of people local to Barrio Logan in the creation of Chicano Park, have been central in the constitution of Chicanxs as "subjects of struggle" (sujetos de lucha), a term that Gutiérrez has used to define people who are formed as a community through their collective action, in so doing producing new forms of cooperation.[91] Through struggle – including the specific and separate daily struggles that each individual person experiences – not only community is formed out of those who are part of the movement, but, as Jenkins has argued,[92] the subjectivity of each individual person is deeply affected. However, while those self-identifying as Chicanxs play an important role throughout this book, it is not the sole identifier of those cited here and interacted with during our research. San Diego is shaped by its proximity to the U.S.-Mexico border as well as being formed by diverse communities, including Latinx, Mexican American, Chicanx, and immigrant communities. In addition, as we will argue in this section, the Chicanx community is characterized by heterogeneity, and this applies to other identifying 'categories' as well. Latinxs form an especially diverse group in terms of race and ethnicity and their legal status; in addition, experiences of discrimination and marginalization are not universal.[93] Notably, experiences, knowledge, and resistance by queer, Indigenous and Black Latinxs in particular are often rendered invisible in narratives and accounts of the Latinx community.[94]

The term Chicanx emerged as a preferred self-identifier over terms such as Latinx or Hispanic, which are themselves of relatively recent origin, entering the lexicon to describe Mexican Americans and some other Latin Americans in the United States in the 1940s.[95] These latter terms imply a Spanish-speaking hemispheric unity, which erases the diversity of identities among Latin Americans as well as speakers of Indigenous languages, so that many reject these terms as administrative impositions. While Spanish speakers in California had for a long time referred to themselves as Hispanos or Hispano-Americans precisely to assert their own racial superiority, Anglo-Americans transferred the term to refer to include all Spanish-speakers of Mexican and Latin American origin.[96] In his book, *The Latino Threat*, Leo Chavez posits that Latinxs are perceived differently from previous immigrants who ultimately became part of the nation – which is particularly evident in media discourse. The dominant narrative, he argues, is that Lat-

---

91 Gutiérrez, What's in a Name?
92 Jenkins, *Extraordinary Conditions*.
93 Chavez-Dueñas et al., Healing Ethno-Racial Trauma, p. 51.
94 García, The Politics of Erased Migrations, p. 3.
95 Gutiérrez/Almaguer, Introduction, p. 2.
96 Gutiérrez, What's in a Name?, pp. 31–38; cf. Aparicio, (Re)Constructing Latinidad; Sánchez, *Homeland*; Gracia, *Hispanic/Latino Identity*.

inxs are "unwilling or incapable of integrating" and are therefore portrayed as "an invading force from south of the border that is bent on reconquering land that was formerly theirs (the U.S. Southwest) and destroying the American way of life."[97] Gutiérrez argues that to be Chicanx is based on an "oppositional consciousness and militant nationalism" that has laid the groundwork for a wider "Latinidad."[98] The narrative around Mexicans in particular (but also Latinxs in general) has been marked by their characterization as "illegal aliens," marking them as "illegitimate members of society undeserving of social benefits, including citizenship."[99]

As Chavez notes, citizenship is a key concept in American culture, which can involve incorporating immigrants into society through a transformation from "other" to "us."[100] However, as we will show in this book, Chicanxs contest this process of incorporation, and instead propose a citizenship that recognizes rather than erases difference. Chavez calls this cultural citizenship.[101] This resonates with Renato Rosaldo's framing of cultural citizenship including the right to be different and yet the same,[102] demanding social justice regardless of origin, phenotype, and status. This is a citizenship which involves active subject-making as a legitimate U.S. citizen in the face of the dominant discourses that portray Latinxs as illegitimate.[103] As we will show, Chicanx subject-making occurs through many practices, including aesthetic ones, that make public statements of belonging, and through which Chicanx subjectivity develops as an oppositional consciousness.

It was in the 1960s that people of Mexican American origin and those with families with a long history in the south-west region of the United States began to identify as Chicanxs. The Chicanx political commitment has its foundation in civil rights movements that protested social inequality, racism, imperialism and violence, including the Vietnam War. It was in the wake of these protests that "Chicano" emerged recognizably as a collective. Anzaldúa describes this process with these words:

> Chicanos did not know we were a people until 1965 [...]. With that recognition, we became aware of our reality and acquired a name and a language (Chicano Spanish) that reflected that reality. Now that we had a name, some of the fragmented pieces began to fall together – who we were, what we were, how we had evolved.[104]

---

97 Chavez, *The Latino Threat*, p. 3.
98 Gutiérrez, What's in a Name?, p. 43.
99 Chavez, *The Latino Threat*, p. 4.
100 Ibid., p. 12.
101 Chavez, *The Latino Threat*, p. 12, citing Flores and Flores.
102 Rosaldo, Cultural Citizenship in San José.
103 See also Flores-Gonzáles, *Citizens but not Americans*.
104 Anzaldúa, *Borderlands/La Frontera*, p. 85.

Gutiérrez writes that Raza communities, including Chicanxs and Boricuas (the self-designation for people from Puerto Rico), took inspiration from the Black Panthers when founding community defense organizations, such as the Brown Berets in the Chicanx case, that embraced the color categories that their grandparents had deliberately avoided.[105] Chicanxs promoted inclusive self-help and supported César Chávez's and Dolores Huerta's unionization campaign for better wages and working conditions for workers of every nationality.[106] The ideals of Chicanismo were expressed spiritually and spatially in the notion of Aztlán, which refers to a presumed region in the Southwest from where the Aztecs are understood to have migrated prior to Spanish colonization (see chapter 4). The *Plan Espiritual de Aztlán*, written in 1969, called for "unity among all racially oppressed groups," community control over local institutions, and the development of institutions that would protect Chicanx civil and human rights and guarantee fair wages.[107] Thus, the Chicano movement particularly championed the rights of the mostly Mexican-born harvesters, but also formed alliances with other groups and was in general open to all segments of society.

To this day, the Chicano movement's heterogeneity has increased through the integration of several generations with very different experiences and challenges. For some members of the younger generation, for example, the "struggle" of the 1970s, which built on the farmworker struggles of the 1930s,[108] is largely alien to their experience. Catherine's conversations with young Chicanxs suggest that, while some come from farmworker families, most relate to agricultural workers' issues in more abstract terms. Nevertheless, they reflect social inequality and institutionalized racism as structural features of their society, even if they themselves are affected by them in different ways than their elders were in their day. Class and educational positionality must similarly be viewed in a differentiated way, since, for example, numerous Chicanx intellectuals are socially mobile in U.S. society.[109]

"Chicanx" is thus not to be understood as a description of a closed community, but rather as individuals of mostly Mexican American descent who form various alliances. Many generally identify with U.S. society – albeit from a critical perspective. Others, like Evelya Rivera, express: "we have experienced alienation, ni de aquí ni de alla, not here nor there. We are not Mexican because this land is no lon-

---

105 Gutiérrez, What's in a Name?, p. 38, p. 41.
106 Acuña, *Occupied America*, pp. 301–310; Gutiérrez, What's in a Name?, p. 41; Garcia, *From the Jaws of Victory.*
107 Gutiérrez, What's in a Name?, p. 43; see chapter 4.
108 Acuña, *Occupied America*.
109 Gutiérrez/Almaguar, Introduction; Sánchez, *Homeland*.

ger Mexico and we are not American because our heritage, our blood, is not American."[110] She explains that it is not through assimilation, but through "reclaiming their history, learning about their heritage, and recognizing their own cultural roots, [that] Chicano/as could become part and parcel of the United States."[111] *Community* is significant not only as a community of solidarity, but, like family, it is a highly politicized and also anti-assimilationist concept in the Chicano movement that promotes resistance to Anglo hegemony. At the same time, with the emphasis on community and family, there is both dissent and consensus with the values of the Anglo-American population, which, however, from Chicanxs' point of view, is more oriented toward individualistic values. When we use the notion of community, therefore, we are not referring to a homogenous group of individuals, but to people connected principally by common values, as well as affect, loyalty and involvement in one another's lives[112] – which, however, does not exclude frictions and conflicts amongst them. This lack of uniformity is also apparent when looking at the Spanish-speaking population segment of the statistical categories Hispanics and Latinos.[113] Here, too, multiple fault lines exist, especially regarding migration from the South. Conflicts between migrants and long-established Mexican Americans fearing for their jobs were reported as early as around 1945 during the *Bracero* program.[114] Even today's middle-class Hispanics who are represented in political office do not necessarily support liberal migration laws,[115] but rather harbor resentment against Mexico as a conservative constituency, advocate tighter controls on the southern border and show themselves to be distinctly "American" in their values.[116]

A borderland city

Our research is situated within what Anzaldúa calls a "borderland"[117]: a space where "two or more cultures edge each other, where people of different races oc-

---

110 Rivera, "Chicanismo," p. 11.
111 Ibid.
112 Brint, Gemeinschaft Revisited.
113 In the U.S. Census, Hispanics are defined as Spanish-speaking persons, regardless of origin or phenotype. Latinxs, on the other hand, include the non-Spanish-speaking population from Latin America, such as from Brazil.
114 De Léon/Griswold del Castillo, *North to Aztlán*, p. 136.
115 Ibid., p. 207.
116 Nevins, *Operation Gatekeeper and Beyond*, p. 104.
117 Anzaldúa, *Borderlands/La Frontera*.

cupy the same territory, where under, lower, middle and upper classes touch, where the space between two individuals shrinks with intimacy."[118] Someone growing up in a borderland may therefore be forced, to place themselves in relation to vague and constantly moving boundaries.[119] It is a space of ambiguity, where national sovereignty is put on display and challenged at once, where people are separated and classified, but also constantly cross borders that are made to keep them apart. Therefore, if "being a settler society structures all American lives,"[120] this is particularly evident in a borderland, where people's lived experiences draw attention to "epistemological and political issues of location."[121]

In the U.S.-Mexico borderland, people who are differently racialized, with different languages, religions and histories occupy overlapping geographical space. This is linked to the history of the U.S.-Mexico border itself. Prior to the Mexican-American War (1846–1848), and the treaty of Guadalupe Hidalgo (1848), San Diego had been part of the Mexican state of California, and had belonged to New Spain before that. The Spanish arrived in the San Diego area and created a settlement (of mostly Indigenous, Mestizo and African people, with few Spanish settlers)[122] in 1769. Following the relocation South of the U.S.-Mexico border, people who had only recently before become Mexicans with the country's independence from Spain in 1821 found themselves caught in a space of ambiguous belonging. Likewise, the imposition of the border separated the Indigenous Tipai-Ipai nation between the U.S. and Mexico. To the South they came to be known "as *Kumiai*, following the Spanish pronunciation, while those in the English-speaking north came to be known as *Kumeyaay.*"[123] The culture of Spanish- and Indigenous language-speakers grated against that of incoming Anglophones who regarded them as Mexicans and Indians. By the time that San Diego had been incorporated into the U.S., those local families who might have been able to trace their residency in the area back more than eighty years would already have experienced one new social and political regime overlaid over another, as well as the concomitant intermixing. The borderland was created by the relocation of the political boundary between the two states, one that would become more physically concrete with time. As we will discuss in chapter 4, this territory overlapped with the presumed ancestral land of Chicanxs, Aztlán. Claiming a pre-colonial heritage and Indigenous belong-

---

118 Ibid., p. 19.
119 See Casaglia, Interpreting the Politics of Borders; Laine/Casaglia, Challenging borders; Van Houtum/Van Naerssen, Bordering, ordering and othering; Scott, *Agenda for Border Studies*.
120 Cattelino, Anthropologies of the United States, p. 275.
121 Gupta/Ferguson, Discipline and practice, p. 39.
122 Hernández, *Coloniality of the US/Mexico border*, p. 6.
123 Ibid.

ing through their connection to Aztlán and their relations with other local Indigenous people, including Kumeyaay people, lends legitimacy to political Chicanismo and its fight for self-determination and sovereignty.[124]

The outcome of the Mexican-American War did not initially lead to demographic change (from 1840–1880, the average migration from Mexico to California was no more than 3,000–5,000),[125] but by the time of the Mexican Revolution (1910–1920) migration had begun to transform the city of San Diego. In particular, the neighborhood of Logan Heights, in the San Diego Bay area, became a heterogeneous, but predominantly Mexican-American neighborhood.[126] Those whose families had lived in the area for generations began to be joined by more recent migrants, some of whom were encouraged to move to the area by employment in the tuna canning industry, and Logan Heights came to be composed of Mexicans and other Raza whose families had been living in the neighborhood since before the Mexican-American war. Due to its strategic location, San Diego also became a permanent location for the U.S. Navy, whose personnel and infrastructure came to dominate the bay area. Following the First World War, in 1919, a naval contract awarded the Navy 98.2 acres of land in San Diego.[127]

Nationally, although around 350,000–500,000 Mexican immigrants to the U.S. were pressured or forced to leave the country during the Great Depression, as a response to labor shortages due to conscription, the Emergency Farm Labor Agreement of 1942, dubbed the *Bracero* program meant that migration from Mexico increased once again.[128] Indeed, through the *Bracero* program, the United States actively encouraged migration.[129] However, the agreement prohibited workers from forming unions and collectively bargaining for increased pay, obliged them to leave the country after harvest season and excluded them from citizenship.[130] This led to a national decline in farm wages from 1942 to 1959.[131] Chacón calls the *Bracero* program the prototype for what became "undocumented migra-

---

124 Rodríguez, *Rethinking the Chicano Movement*.
125 Gutiérrez, Historic Overview of Latino Immigration, p. 109.
126 Migration from Mexico to California as a whole was only on average 3,000–5,000 people per decade from 1840–1890, but this number increased significantly in the final decade of the 19th century. By 1900 around 100,000 Mexicans had migrated, and this doubled to 220,000 when the Mexican Revolution began in 1910 and doubled again to 478,000 by the end of the Mexican Revolution in 1920, see Gutiérrez, What's in a Name?, p. 109.
127 Galaviz, *Expressions of Membership and Belonging*, pp. 25 f.; Norris, Growth and Change.
128 Gutiérrez, Historic Overview of Latino Immigration, pp. 109 f.; McCaughan, *The Border Crossed Us*.
129 Saldívar, Border thinking, p. 275.
130 Chacón, *The Border Crossed Us*, p. 70.
131 Ibid.

tion."[132] While between 1942 and 1946 4.6 million Mexicans became braceros, those excluded from the program became the first unauthorized laborers. Unauthorized migration of those deemed ineligible for the *Bracero* program continued alongside the growing legal migration of Mexicans to the U.S. from the 1940s to 1960s.[133] A large number of people that had entered the U.S. legally did not return to Mexico after the program ended and thus went from being legal workers to "illegal aliens." At the same time, the U.S. feared the intrusion of enemy agents across the southern border. In this context, the issue of border security[134] was declared a national security problem.[135]

Until the present day, migration policies have fluctuated with legal measures to stem its flow, and measures to legalize the status of long-time unauthorized workers in the U.S. would be accompanied by the sanctioning of contemporary unauthorized workers. Towards the end of the twentieth century, for example, the 1986 Immigration Reform and Control Act (IRCA) conferred citizenship on 2.7 million unauthorized workers while also criminalizing labor migration.[136] At the same time the IRCA doubled the budget for the Immigration and Naturalization Service (INS) from $577m to $1.5 billion between 1986 and 1995 and placed greater emphasis on stopping migrants at the border itself, for example through the militarization of the border in Operation Blockade (1993) and Operation Gatekeeper (1994).

The border, however, is highly dynamic and is made more permeable or more hermetically sealed off as needed, accompanied by corresponding discourses that are fueled by the media. While the armament and militarization of the border was advanced in 1994 during Operation Gatekeeper, the North American Free Trade Agreement (NAFTA) came into force at the same time, establishing a close economic link across the border.[137] Following 9/11, the PATRIOT Act was passed to allow immigration agencies to deport anyone deemed a security threat without a hearing. More recently, the creation of the migration enforcement agency ICE and the expansion of the border wall have made life more difficult both for migrants attempting to cross the border, and those wishing to remain in the U.S.[138]

San Diego was chosen as a research site because of the nature of its history and contemporary character as a borderland city, and its present-day demographics, with a significant "Hispanic" (the category used in the census) population,

---

132 Ibid.
133 Gutiérrez, Historic Overview of Latino Immigration, p. 110.
134 Arfsten, *The Minuteman*, pp. 63 f.
135 Nevins, *Operation Gatekeeper and Beyond*, p. 38.
136 Chacón, *The Border Crossed Us*, p. 178.
137 Arfsten, *The Minuteman*, pp. 66 ff.; cf. Hernández, *Coloniality of the US/Mexico border*, p. 182.
138 Chacón, *The Border Crossed Us*, pp. 188, 193 f.

Map 1: Red represents white, blue represents Black, green represents Asian, orange represents Hispanic, yellow represents other. Each dot is 25 people. Data is from census 2010.

alongside a majority "white" population. The city of San Diego has a population of 1.387 million people as of the census of April 2020.[139] 30.1 % of the San Diegan population is identified in the census as "Hispanic," with "White Hispanic" being the second biggest category in the census adding up to 19.7 % (see chart 1, p. 25). 88.6 % of residents of San Diego, California are U.S. citizens.[140] As map 1 shows, the pop-

---

139 Source: https://www.census.gov/quickfacts/fact/table/sandiegocitycalifornia/PST045221.
140 Source: https://datausa.io/profile/geo/san-diego-ca. The census categories are problematic from an anthropological point of view, not least because although they ask respondents to self-identify according to a particular category, the categories themselves are already given, and arbitrary. The census categories include Hispanic, for example, but not Latinx or Chicanx (and as chart 2, p. 26 shows, "Hispanic" encompasses a broad range of identities). There are also many administrative

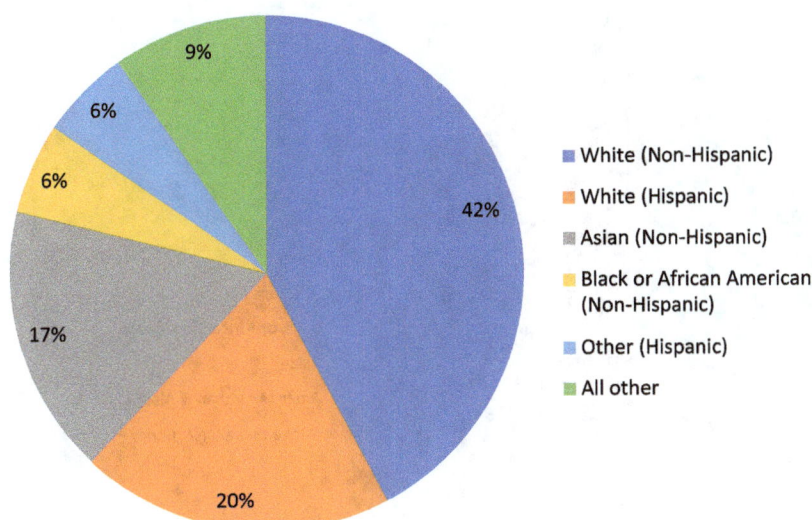

Chart 1: Five largest ethnic groups in San Diego according to census 2020.

ulation categorized as Hispanic tends to be concentrated in the neighborhoods south of downtown and close to the border, while the white population is spread around neighborhoods along the northern beaches and suburbs of San Diego.

The San Diego-Tijuana region has been framed as a "hybrid" border space, as a contested identitarian space but also as a circuit of exchange and crossovers.[141] Pablo Vila argues that it is differences rather than hybridity that is a prominent feature of the border region.[142] According to Herzog and Sohn, this border zone is shaped by debordering and rebordering mechanisms – debordering emphasizing interaction and flows and rebordering focusing on hardening the border for security reasons.[143] An initially non-existent border became first a porous and over time increasingly impenetrable one.[144] While historically existing in the "shadow of San Diego" and functioning as a city to fulfill the needs of its northern twin neighbor,[145] Tijuana's development is closely linked to the debordering and

---

hurdles to overcome in order to be included in the census in the first place, thus limiting its inclusivity.
141 Canclini, *Hybrid Cultures*; Kun/Montezemolo, The Factory of Dreams.
142 Vila, The Polysemy of the Label "Mexican."
143 Herzog/Sohn, The co-mingling of bordering dynamics, p. 184.
144 St John, *Line in the Sand*.
145 Sparrow, San Diego–Tijuana, p. 76.

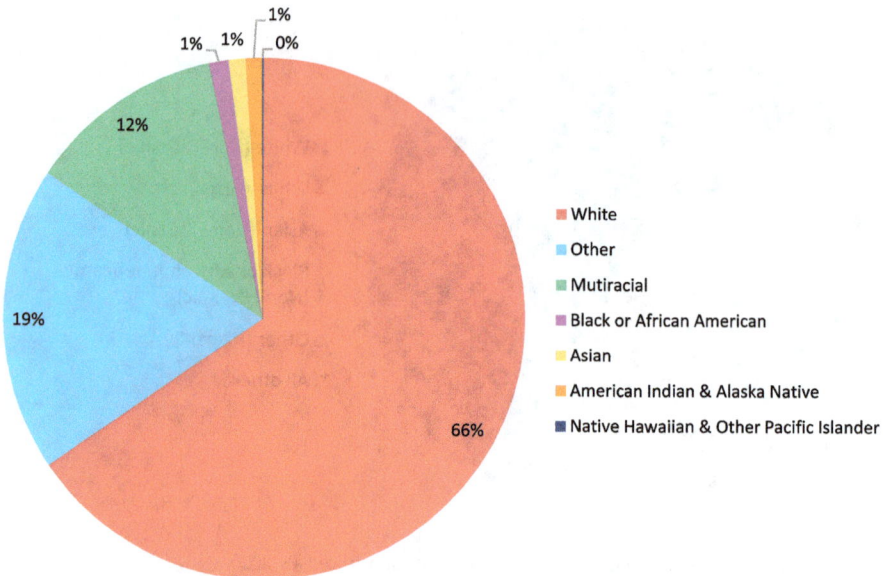

Chart 2: Hispanic population in San Diego according to census 2020.

rebordering forces.[146] Crossing the border from Tijuana often happens daily and more permanently for work purposes,[147] as well as for more affluent families for their children's education and shopping in San Diego.[148] Border crossing from north to south is often motivated by the desire for activities, goods and services that are cheaper in Tijuana than in San Diego (or even prohibited there), such as prostitution, bull and cock fighting or gambling. Thus, while Tijuana is attractive for some with an almost romantic connotation as the "happiest place on earth," it is frightening for others, who describe it as a "drug capital."[149] Tijuana is also an environment that offers low-cost employees, as well as lesser environmental and worker regulations for companies from the North, along with being a tourist destination for the U.S.[150] It is common, for example, for people in San Diego to see doctors and dentists in Tijuana at a fraction of the cost that they would pay at home. However, there are also many white San Diegans who have never been to Tijuana. The border is often mentioned as a source of fear (citing drug cartels

---

146 Herzog/Sohn, The co-mingling of bordering dynamics, p. 194.
147 Yeh, *Passing*.
148 Sparrow, San Diego–Tijuana, p. 79.
149 Kun/Montezemolo, The Factory of Dreams, p. 3.
150 Sparrow, San Diego–Tijuana, p. 76.

Map 2: Map of the main research locations in the city of San Diego.

and violence), when people remark: "We are only 15 minutes from the border," as you can read on the motorways. Tijuana is known for its high levels of crime and violence and "has broken all of the most brutal records of national violence,"[151] however, while poverty is commonly attributed to the southern side of the border, more people are unhoused in San Diego than in Tijuana.[152] The twin cities San Diego-Tijuana are thus heavily shaped by the border and border practices and while their relationship is uneasy, the economic interdependence between the two has increased since the 1970s.[153] In this book, we focus on the U.S. side of the border.

Map 2 shows our main research locations in San Diego. Mostly, our research took place in Logan Heights and Barrio Logan, as well as the Centro Cultural de la Raza, a Chicano Cultural Community Center located in Balboa Park. Another important site was Chicano Park in Barrio Logan, which is a community park created in an act of resistance to the building of Coronado Bridge in the late 1960s. Chicano Park's famous murals, which were painted on the pillars of the bridge, narrate Chicanx history and identity, as discussed in chapter 3 and 4. In chapter 5 we extend our description of San Diego by looking at different neighborhoods in comparison.

---

151 Aviña Cerecer, The Dispossessed of Necropolitics, p. 7.
152 Ibid., p. 4.
153 Sparrow, San Diego–Tijuana, p. 77.

We also reference these sites' virtual existence in the digital sphere throughout the book. In chapter 6, however, social media becomes the main focus – as a research site and as a space where watchfulness and "borderlands" expand into.

In Logan Heights, from the 1930s onwards, the Mexican-American character of the neighborhood would be further reinforced through the practice of redlining, a discriminatory practice of categorizing neighborhoods according to the supposed risk that they posed to insure.[154] This practice developed following the 1934 Housing Act, part of President Roosevelt's New Deal, through which the Federal Government reshaped housing finance to stabilize housing markets and support lenders following the Great Depression.[155] Although not officially a categorization according to race, those neighborhoods deemed as safest to insure were always those that were exclusively white, as opposed to mixed ethnicity or Black or Hispanic neighborhoods.[156] Redlining came to determine people's abilities to get loans and mortgages (and therefore the type of housing and areas they could live in), and public services provided. As Valenzuela-Levi shows in relation to the provision of internet services in Santiago Chile,[157] redlining can mean that neighborhoods are not taken into account equally in the provision of necessary infrastructure and services (so that those excluded have to take matters into their own hands to access them). While the 1958 Fair Housing Act was supposed to encourage fair housing opportunities regardless of race, religion or national origin, by then the practice of redlining had set up the conditions for largely ethnically homogeneous neighborhoods across the U.S. Redlining defined the areas where Hispanic people could live: in San Diego these were Logan Heights, Lemon Grove where Mexicans picked Lemons, Little Italy, and a tiny spot in La Jolla (where maids lived). Though Spanish-speaking people had long lived in these areas, they became second-class citizens.

Neighborhood organizations in Barrio Logan draw parallels between current discriminatory practices and displacements that people living there have faced since the U.S. won the war with Mexico, but also since colonization of the area by the Spanish. Logan voices are hardly taken into account in local decision-making, such as the construction of the Coronado Bridge, which divided Logan Heights and created the neighborhood of Barrio Logan in 1969. At a monthly meeting of the Unión del Barrio's *Noche de Resistencia in Chicano Park* in February 2020, one of the presenters showed the group a map of early Logan Heights (which included

---

154 Areas were graded A (least risky) to D (riskiest), and the term redlining was thought to derive from the red shade demarcating the D neighborhoods (Aaronson/Hartley/Mazumder, 'Redlining' Maps, p. 7).
155 Aaronson/Hartley/Mazumder, 'Redlining' Maps, p. 34.
156 Rothstein, *The Colour of Law.*
157 Valenzuela-Levi, "Somos Zona Roja."

Barrio Logan) as a redlined district. Discriminatory urban planning practices transformed the neighborhood officially from a residential to a mixed-use (residential and industrial) zone, and as a consequence, between 1940 and 1970, the population of Barrio Logan would drop from a high of 20,000 to 5,000.[158] A series of rezoning ordinances in 1957 led to further heavy industry entering the neighborhood, and the construction of the I-5 freeway to the community losing their previous easy access to the bay and the beach.[159] Because of their lack of political capital, the losses of the Mexican Americans displaced from the neighborhood were not taken into account prior to its construction. Current residents of Barrio Logan continue to push back against changes to the character of the neighborhood. We will argue in chapters 5 and 6 that the struggle that the 1960s Chicanx movement engaged in against the construction of intrusive infrastructure, finds echoes in current struggles against the gentrification and "gentefication" (a Chicanx slang word for the replacement of working-class Chicanx culture with its middle-class equivalent) of Barrio Logan, which has seen the Chicanx struggle become commodified through the appearance of cafés, art galleries and other shops that are drawing tourists to the neighborhood.[160] As the population of Barrio Logan grows more diverse, its struggles become more fragmented, often pursuing separate, sometimes competing, or conflicting goals.

Coloniality in San Diego

By creating ethnically homogeneous neighborhoods, practices such as redlining are a clear example of what Peruvian sociologist Aníbal Quijano has called "the coloniality of power."[161] That is, the formal or informal replication of colonial practices of discrimination and inequality in post-colonial societies over new institutional bases. Roberto Hernández has referred to the "socially produced and contested space" of the border as "coloniality incarnate."[162] The effects of the border are felt by Chicanxs through the U.S. cultural imagery of the border itself protecting Americans from racialized undesirables[163] who are a "threat" to U.S. society.[164] The border is also felt through legal practices that do not separate people at the

---

158 Rosen/Fisher, Chicano Park.
159 Galaviz, *Expressions of Membership and Belonging*, p. 30.
160 Delgado/Swanson, Gentefication in the Barrio.
161 Quijano, Coloniality of Power.
162 Hernández, *Coloniality of the US/Mexico Border*, p. 5.
163 Ibid., p. 44.
164 Chavez, *The Latino Threat*.

physical border, but create "borders of belonging"[165] within the United States itself. Coloniality is felt in the way that decisions made at city level are more likely to adversely affect Chicanx neighborhoods such as Barrio Logan. This has resulted, for example, in Barrio Logan becoming the most polluted area of the city (see chapter 5).

Discriminatory policies regarding pollution reinforce assumptions about whose voice does and does not count locally, whose voices are able to be heard, and taken into account, and which groups "have no part."[166] The dominant social paradigm in the United States privileges Anglo-American perspectives on what it means to be a citizen, how one is expected to look, behave and to speak. This allocation of ways of being, doing and saying is what Rancière calls the "police." The "police" order of the United States is one in which coloniality is taken for granted as a "symbolic construction of the social".[167] Rancière calls the division of the world into parts, the "distribution of the sensible." This finds its physical expression in infrastructure projects such as the Coronado Bridge, built through Barrio Logan, which will be discussed in chapter 3, which makes coloniality manifest by the very fact that the people of Barrio Logan were given no say in its construction.[168] Rancière contrasts the concept of the "police" with "politics," which makes "what was unseen visible."[169] In this ethnography we highlight how the Chicanx organizations that emerged in San Diego from the 1960s have shone a light on the daily structural inequalities that they face and how they have challenged them.

The coloniality of the borderland condition requires Mexican Americans and Chicanxs and other racialized people to justify their own citizenship. This is because the structural inequality that many people living in San Diego must deal with includes the Anglo-American assumption that they are (possibly illegal) migrants with less right to live in the United States. This sentiment is made manifest, for example, when Mexican American protesters are told "Go back to your country!" by Anglo-American political opponents, or when a Mexican American cleaner is expected to work under the table at a low cost with the assumption that she is living in Tijuana and not paying U.S. taxes. The desire to avoid such suspicions, we argue, results in a watchfulness of being, which was particularly accentuated by the political climate during which the main part of this research took place between February and December 2020. This was the period leading up to the end of Donald Trump's period in office as U.S. president, whose rhetoric and policies,

---

165 Castañeda, *Borders of Belonging.*
166 Rancière, *Dissensus.*
167 Ibid., p. 36.
168 Alderman/Whittaker, A Bridge that Divides.
169 Rancière, *Dissensus*, p. 36.

i.e. emphasizing the need for a "big beautiful" border wall, constantly stoked the fears and suspicions of Anglo-Americans towards migrants, and Mexicans in particular.

This politicized debate influences the subject positions of the Spanish-speaking population and largely constructs them as "foreign" and as not belonging per se to the imagined nation. The perceived intrusion of foreign bodies into U.S. society is countered with biopolitical measures of disciplining and governance in the sense of Foucault, which manifests itself in sophisticated surveillance apparatuses and protective walls from the state side, but also in specific practices exercised by the non-state side. This includes, for example, vigilante groups that have formed in the border region along the lines of the patriotic militias of the War of Independence and, as Minutemen, seek to protect the U.S.A. from the invaders. Even if these developments are being discussed particularly virulently in the aftermath of Donald Trump's presidency, they point to a certain historical continuity.[170]

This increased the need for Latinxs and Chicanxs to be watchful in their everyday behavior, particularly in encounters with representatives of state agencies, such as the police, which shape them as vigilant subjects. However, suspicion can be discerned differently throughout the city: in the middle-class neighborhoods to the north, defined primarily as white, the border loses its salience and plays only a minor role in the daily lives of residents there. In these districts, where white, middle-class neighborhoods dominate, residents of Mexican origin are more conspicuous than in southern districts, which tend to be populated by non-whites – as is the case in Barrio Logan.

The biopolitics of the state is combined with a necropolitics that finds expression in various forms of health injustice, including elevated levels of asthma and cancer in Barrio Logan as a direct result of the heavy industry and infrastructure that city authorities have chosen to place within the neighborhood. Necropolitics is defined by Achille Mbembe as "the power and capacity to dictate who may live and who must die."[171] This can include both the active decision to place polluting industries and infrastructure in locations where they will affect specific communities, and the inaction of regulating authorities who do not sufficiently and equally consider the lives of different sets of people. As we argue in chapters 5 and 6, the effects of being on the receiving end of unequal political decisions are felt physically through health problems that are a direct result of the low-quality air that people have to breathe and other discriminatory conditions and the way that the struggle against coloniality becomes itself imprinted physically on their bodies. However,

---

170 Arfsten, Auf der Jagd; Shapira, *Waiting for Jose*.
171 Mbembe, Necropolitics, p. 11.

their continuing survivance,[172] involving contemporary resistance against the designation of the neighborhood as a mixed-use zone (where heavy industry can operate alongside residential homes) appears to be bearing some fruit. In chapter 6, we look at how these health issues as well as the underlying discrimination is addressed through healing practices that are inextricably linked to a call for social justice. In this part of the book we argue that reclaiming ancestral forms of healing is not only a way towards the decolonization of the self, but also a source of empowerment and social justice activism for those who are especially affected by discrimination in the U.S. health care system, as well as throughout all of society.

Structure of the book

We trace the formation of vigilant subjecthood among migrantized San Diegans in the following five chapters:

In chapter 2 we outline the methodological approaches we took during different stages of research. We describe how the principal fieldworker in the group, Catherine Whittaker, entered the field, and then how we approached a later period of joint fieldwork in San Diego. We reflect on the particular challenges related to the place and time in which fieldwork was conducted. As a borderland city, where, we argue, watchfulness is central to people's everyday lives and subjectivities, some of the people that Catherine and later the whole team met in San Diego are suspicious towards researchers, whether local or from further afield. We examine the challenges that had to be confronted in overcoming this watchfulness, as well as describing some of the relationships with people who were more open to assisting and collaborating with our research from the beginning. Conducting field research during 2020 and 2021 was also challenging because of the COVID-19 pandemic. We describe how the fieldwork evolved to take into account the resulting restrictions, but also to include unexpected events such as the Black Lives Matter protests.

In chapter 3 we take the construction of the Coronado Bridge in 1969 as an inflection point in the history of the neighborhood of Logan Heights and what became Barrio Logan that was significant in the development of Chicanx subjectivity in San Diego. We look at how local people responded to the construction of this unwanted infrastructure by establishing a community space below the bridge, Chicano Park. Chicanx subjectivity developed partly through the struggle to assert control over this space, and because the murals that were painted on the supports of

---

172 Vizenor/Lee, *Postindian Conversations*, p. 82.

the bridge reflect the local Chicanx vision of their history, with both local and wider Latin American aspects to it.

In chapter 4 we continue to examine Chicano Park as a site of subject-formation and quotidian watchfulness.[173] Through murals that depict Aztlán, the Chicanx spiritual homeland, the temporal aspect of the Chicanx decolonial struggle is emphasized. We describe Chicano Park as the center of this homeland. Through the example of Joaquín, we show that Chicano Park is not a safe space for Chicanxs by default, but one that the community has to struggle for continuously and individuals have to move through watchfully.

Chapter 5 looks at the way that coloniality becomes imprinted on the body itself. The heavy industry that has developed in Barrio Logan since the early twentieth century, combined with the construction of polluting infrastructure described in chapter 3 has directly resulted in Barrio Logan becoming one of the neighborhoods with the worst air quality in the city and California as a whole. As a consequence, residents of Barrio Logan have a higher-than-average incidence of health problems such as cancer and asthma. We reflect on the intersectional nature of health and how this is reflected in experiences of the pandemic and of gentrification among people in different neighborhoods across San Diego. Chicanx activists therefore express their survivance through environmental justice campaigns that highlight continuing coloniality.

Chapter 6 examines how watchfulness of the self and others can extend into the digital sphere. Here, a group of Black, Indigenous, and People of Color – self-identifying as *brujxs* – combines their healing practices with a call for social justice. Watchfulness is shown through practices that are understood as part of healing from injustice: one's constant self-reflection as "shadow work" as well as publicly "calling out" those who are perpetuating harm. In this chapter, we show the role that social media plays for watchfulness against social injustice, especially for racialized and disadvantaged people.

In our conclusion, we trace how we arrived at our main arguments and acknowledge the limitations of our work. In particular, we highlight the book's contribution to the theoretical literature alongside which we place our work. We suggest that our analysis of Chicanx watchfulness through the concept of being "trucha" advances studies of vigilance and subjectivation by highlighting how watchfulness incorporates resistance – in the Chicanx case, resistance to coloniality. This alert resistance in terms of being *trucha*, we argue, has powerful subject-forming effects. To be watchful – *trucha* – is a significant aspect of what it means to be Chicanx.

---

173 Amit, Rethinking Anthropological Perspectives.

Chapter 2
A Hall of Mirrors: Watchfulness as Ethnographic Method

Ellos son Nosotros / Nosotros somos Ellos / Ellos están luchando / Y tú Qué?
They are Us / We are Them / They are fighting / What about You?

It was November 2019 on her first visit to San Diego, California, and a 31-year-old brunette white woman was looking back at Catherine: her own mirror image. She was in front of an art installation at the Centro Cultural de la Raza, a 50-year-old key Chicanx institution in the heart of the border city. The mirror between the Spanish and the English poems invited visitors, the majority of whom were Chicanxs and Latinxs, to reflect about their identity and their solidarity with unauthorized migrants to the U.S., who often risked their lives in attempting to cross the border, only to get caught and imprisoned in detention centers on the other side. The Centro collaborated with artists and activists from the Otay Mesa Detention Resistance to raise awareness of this ongoing humanitarian crisis and to recruit more supporters for the cause.

As the sociologist Pablo Vila argues, "the border offers multiple mirrors from which to view oneself and others."[1] For Chicanxs and other people of Mexican descent, living in the borderlands means constantly being confronted with narratives about their ethnicity and their nationality, and thus having to reflect on their identity and how this positions them towards Anglo-Americans and Mexican nationals.[2] Anthropologists, too, are often border people: "between places, between identities, between languages, between cultures, between longing and illusions, one foot in the academy and one foot out."[3] For Catherine, the mirror was also a reminder that, as anthropologists, "not only do we observe, ask questions and collect data, but we are also simultaneously the objects of sometimes secret – and very often quite obvious – observation and surveillance"[4] – in this case, of herself. Who else is watching? What does it mean to ethnographically observe people who are attentively observing themselves and others? And how should anthropologists deal with demands for allyship, or even complicity in sites of anticolonial struggle?

---

1 Vila, The Polysemy of the Label "Mexican," p. 138.
2 Ibid.
3 Behar, *The Vulnerable Observer*, p. 162.
4 Sökefeld/Strasser, Under suspicious eyes, p. 161.

Open Access. © 2023 the author(s), published by De Gruyter. This work is licensed under the Creative Commons Attribution 4.0 International License. https://doi.org/10.1515/9783110985573-003

As we will explain in more detail in the following chapters of this book, among Chicanxs and many other racialized people in San Diego, their vigilance stems from an awareness of being watched and being under a constant, oblique threat of being disadvantaged, or even attacked by white Anglo-Americans on the basis of being read as "immigrants" because of their phenotype. In contexts of ongoing coloniality, groups identified with the colonized and the colonizers watch each other closely, anticipating potential power moves and threats to their existence. However, less powerful, marginalized citizens are more vulnerable and therefore practice vigilance more than the dominant population group. Nowhere is this more clearly visible than in highly surveilled borderlands, such as Southern California, where the so-called Global North meets the Global South. Between the COVID-19 pandemic, Black Lives Matter protests, and the U.S. presidential elections, the unique circumstances of 2020 added an additional layer of watchfulness to everyday life.

Anthropological field research generally seeks to understand complex cultural meanings via "participant observation," which means "going out and getting close to the activities and everyday experiences of other people,"[5] a method that the Chicano anthropologist Renato Rosaldo famously described as "deep hanging out."[6] However, "one is never either simply a participant nor simply an observer," as "observing means that the participation itself is objectified by the fieldwork, and indeed, this is why reflexivity is an essential component of any fieldwork."[7] Following the theoretical traditions of symbolic interaction, ethnomethodology, and "situated knowledges,"[8] Catherine attempted to become as immersed as possible in our research subjects' lives within the time available, which "precludes conducting field research as a detached, passive observer" and "inevitably entails some degree of resocialization."[9] In other words, she was changed by the research experience. While the classic cultural relativist approach in anthropology assumes a difference between "self" and "other," "situated knowledges" do not: "The knowing self is partial in all its guises, never finished, whole, simply there and original; it is always constructed and stitched together imperfectly, and therefore able to join with another, to see together without claiming to be another."[10] The effect is to complicate dominant narratives of self and other, as this approach emphasizes the fluid and dynamic, always relational and historically contingent processes of cul-

---

5 Emerson/Fretz/Shaw, *Writing Ethnographic Fieldnotes*, p. 2.
6 Cited in Clifford, Anthropology and/as Travel, p. 5.
7 Fontein, Doing research, p. 75.
8 Haraway, Situated Knowledges.
9 Emerson/Fretz/Shaw, *Writing Ethnographic Fieldnotes*, p. 3.
10 Ibid., p. 586.

ture-making, allowing for the redefinition of identities within it. Accordingly, through long-term exposure over a period of ten months (between February and December 2020), Catherine herself began adopting some of the guardedness of her interlocutors – a point we will return to later. Some anthropologists who are more influenced by the scientific method seek to avoid this kind of "contamination,"[11] yet others view this as advantageous to "understanding the more subtle, implicit underlying assumptions that are often not readily accessible through observation or interview methods alone."[12] This immersive methodological approach equally exposes ethnographers to being observed, the extent and consequences of which may not be known to them.[13] In San Diego, racialized political tensions exposed our team to a high level of scrutiny and suspicion. Hence, in studying watchfulness among those falsely labeled as "immigrants," our team became an object of vigilance for our interlocutors. As a team, Carolin, Catherine, Eveline, and Jonathan also mutually monitored each other's work, which added additional levels of reflexivity to our research. If anthropology seeks to hold up a mirror to the world, our research involved a veritable hall of mirrors.

In this chapter, we will discuss the dynamics involved in these multiple forms of mutual observation and how this shaped our methodological approach. We will begin by discussing the role of vigilance in anthropological research. This will lead us to explain our criteria for identifying our research subjects and for framing our research question. Then we will turn to Catherine's arrival in San Diego and her first experiences of being watched suspiciously, not only when entering the gentrifying Chicanx and Latinx neighborhood of Barrio Logan for her research, but also at home with a Republican Evangelical couple in the diverse lower middle-class suburban neighborhood of Clairemont. In the following section, we will address how we navigated the particular circumstances of the pandemic by expanding Catherine's stay in San Diego, while relying more strongly on digital methods, intimate relationships with a small number of interlocutors, and auto-ethnography, as well as participant observation in outdoor settings, such as protest marches and community garden volunteering. Next, we will question the common assumption that gaining access to a fieldsite and developing rapport with potential research subjects is a linear process, in which trust steadily grows. In the context studied, trust was often fragile between people in general, and had seemingly unpredictable ups and downs. Conflict either led to permanent rupture of a relationship, or its (temporary) strengthening. Developing intimate friendships, weathering inter-

---

11 E.g. Beer, Systematische Beobachtung, p. 174.
12 Emerson/Fretz/Shaw, *Writing Ethnographic Fieldnotes*, p. 4.
13 Verdery, Observers Observed.

personal conflicts, participating in local struggles, and eclectic secondary research helped Catherine to develop a certain wokeness that allowed her to anticipate interlocutors' concerns and expectations more accurately. This became evident in our group fieldwork, which also involved a certain degree of mutual watching between ourselves and methodological trial-and-error. Further, we discuss how group fieldwork carried over into a collaborative writing process in which each author became first author of part of the book but contributed to all chapters, reflecting on the challenges and benefits of this process. In our conclusions, we will argue that decolonizing research not only requires being a good ally, but also accepting the discomfort of constantly being monitored and held accountable, and potentially refused, by both one's interlocutors and one's colleagues both during the fieldwork phase and after.

Suspicious eyes meet Anthropology's colonial gaze

Like the borderlands and Chicanxs, many anthropologists are liminally positioned between the so-called Global North and Global South, both in a geographic sense, in terms of where they conduct their research, and in an epistemic sense, as their concepts and theories intertwine elements of thought from various parts of the world.[14] Their frequent travel to distant locations, some of them shunned by tourists, persistent questioning, and constant note-taking have often rendered anthropologists suspect, both among the people they study and the local authorities.[15] Even anthropologists are often suspicious of other anthropologists' intentions in light of Anthropology's origins as a "colonial handmaiden" and some American anthropologists' ongoing work for the CIA and the military.[16] In present-day San Diego County, for example, anthropologists work as "cultural instructors" at the Coronado Navy base. Some Native American and Latinx anthropologists have recently called for "ethnographic refusals" to circumscribe identities in a way that might further entrench colonial gender binaries and racial hierarchies in our scholarly research and writing.[17] Such refusals cannot be taken for granted, as there are lively ongoing debates surrounding the decolonization of representation, cultural ownership, and related ethical issues in ethnography.[18]

---

14 Comaroff/Comaroff, *Theory from the south.*
15 Sökefeld/Strasser, Under suspicious eyes.
16 Price, *Weaponizing Anthropology.*
17 Simpson, *Mohawk Interruptus*; Chávez/Pérez, *Ethnographic Refusals.*
18 Alonso Bejarano et al., *Decolonizing ethnography*; Little/Rees, Participatory Research.

Accordingly, anthropologists are often quick to admit that research subjects have every reason to be skeptical of their intentions. In the highly politicized settler-colonial borderland context of San Diego, many research subjects from marginalized backgrounds were very guarded when interacting with white foreign researchers like us. Our team had expected this. Yet we were surprised to find that even scholars who are long-term, active members of the local Chicanx community were met with suspicion and resentment. Two Chicanx Studies professors from different local universities explained that they initially had difficulty finding support for their projects, as community members feared being exploited, while more privileged, educated and often more light-skinned Chicanx scholars built their career off the back of their life stories. One recalled being asked, "Why are you Chicano Studies professors interested in our work only now?" He explained that he had been out of town to study, but that he was going to be present in community organizing efforts now. Ever since, he has worked hard to be transparent, to show commitment and to prove that he is not there to enrich himself. The key, he said, lies in being in continuous conversation with the community.

Part of the problem is that the community receives many researchers' requests, from which they seldom benefit. In February 2020, at the beginning of her fieldwork, Catherine attended a Chicano Park Steering Committee (CPSC) meeting. As it had rained the previous day, it was cold at the open-air meeting place, the pyramid-shaped *kiosko* under Coronado Bridge. The core committee members were sitting on foldable chairs, while the rest were standing around them in a circle. There was a constant background noise of cars and trucks whooshing over everyone's heads, which made it difficult to hear one another. Normally, the steering committee would have met indoors, but their usual host was offended because her shawl had allegedly disappeared after the last meeting. An elder, Jacob, pointed out that this was bad, not only because of the emotional value the shawl had for this person, but also because this tarnished the reputation the committee has worked so hard to establish. He said that he did not want to accuse anyone of stealing, but if someone had accidentally picked up the shawl, thinking it was theirs, he asked them to please return it. Even though no accusations were made, this scene suggested that even committee members do not necessarily fully trust each other.

At that same meeting, a Fulbright scholar from India received harsh responses to her request for interviews. One committee member corrected her use of "Mexican American," explaining that the people present were "Chicanos" and that this was not the same thing. Both a female elder and Jacob additionally expressed annoyance at having researchers come and study them all the time, take knowledge, build their careers, enrich themselves, and give nothing back. "I have it up to here," the woman said. The first thing you need to do, is explain what you will do for the community as a researcher, they said. How will your research benefit the community? Later, one committee member expressed that he ignored the scholar's interview request because he is too busy and she "doesn't understand how things work here." You could not just parachute in and expect people to want to work with you. There was another request by a person who was not present and who wanted to shoot a documentary in Chicano Park. Not having explained exactly what this is about, the committee declined. They mentioned still being upset at having recently discovered that a car commercial was being shot in the Park, but they managed to protest and stop it, for which they thanked the person who reported it.

Catherine committed a faux pas at the meeting by accidentally speaking out of turn, for which she was told off by the president of the committee. Inspired by her previous experiences of working with creative writers in Scotland, Catherine said that she was looking for collaborators to organize a series of writing workshops to help anyone in the area who is interested in developing their voices. At the end of the workshops, there could be a closing event where the participants could present their work. She was later advised that she should not have said anything and instead have attended a few times before making proposals, which need to be well thought out. People needed to see that she was there to stay, looking to learn in a humble manner, and fully participate in community organizing work as a volunteer. However, at the end of the meeting, Catherine was given an opportunity to speak again. She apologized and briefly repeated her idea. One of the women seemed confused and asked, "so what is your request?" Another woman said, "she's not requesting anything, she's just making an invitation." Catherine confirmed this. This scene suggested that the CPSC was so overwhelmed by the many requests it receives that the notion that someone might be offering something, rather than taking from them, was unexpected to the point of causing confusion.

After the meeting, Jacob waved Catherine over to himself and asked her to repeat her idea because he was somewhat hard of hearing. He made some suggestions as to whom she could speak to and said that she should not let herself feel discouraged. If she continued showing up, allowing people to get used to her and get to know her, this would make collaborations easier. "This community has been through a lot, so they don't trust easily," he explained. She expressed understanding and volunteered some information about her background, explaining her immigrant, mixed identity as the German-born daughter of an Italian and an Irishman. "You have all those different origins?" He seemed impressed and said, "in Spanish, we'd call you a *mestiza*." Catherine smiled and mentioned that she had lived in Mexico. Establishing this small commonality destabilized the imagined dichotomy between "us" as anthropologists versus "them" as research subjects. Somewhat surprised, Jacob noddingly approved and suggested that Catherine should volunteer at the 50th anniversary of Chicano Park's founding in April.

As the example of the meeting illustrates, finding willing interview partners and gaining access to relevant events was a challenge for our project, as we did not have longstanding relationships of trust and mutual help to build on. Yet beyond the issue of rapport, to which we will return to later, there was also a conceptual problem to solve with respect to our initial research question: If our aim was to study the "watchfulness of those mistaken for migrants," which was our project title in the beginning of this research, how could we identify people who are mistaken for migrants without ourselves enacting the racist gaze?

Learning whom to ask

Learning how to ask questions in a way that reveals the information one is actually looking for is a complex matter. To begin with, choosing the appropriate language is important: "it is often a question of respect, of breaking barriers, of gaining access, and of building rapport."[19] For Catherine, introducing herself in Spanish to people when it was clear that this was their preferred language proved an important asset. For example, while the Chicanx organization Unión del Barrio had the reputation of not being easy to approach, Catherine did not have this impression herself. Speaking Spanish with the group and being able to mention her previous experiences in Mexico as well as her connections to well-respected scholars at UCSD provided some common ground and helped her to be accepted by the members present and invited to other events of theirs.

As the linguistic anthropologist Charles Briggs argued on the basis of his experiences of working with Spanish-speakers in northern New Mexico, beyond the language, ethnographers need to spend considerable time learning key local concepts and forms of knowledge acquisitions, before they are in a position to ask ethnographic questions effectively, in a way that is conducive to both mutual understanding and respect.[20] Following this logic, secondary research was highly important both before and during fieldwork, in order to become conversant in local concepts and narratives, and to avoid exhausting our interlocutors' patience by asking trivial questions, instead showing that we had done our homework and came well prepared.[21] Catherine regarded prior acquisition and utilization of local knowledge as a question of respect and believes that it helped her to build trust with community members. Thus, apart from reading relevant academic liter-

---

19 Fontein, Doing research, p. 84.
20 Briggs, *Learning How to Ask*.
21 See also Thin, Importance of Secondary Research.

ature on the San Diegan context and following the news on San Diegan Latinxs, Catherine sought to deepen her level of cultural immersion by attending meetings of Chicanx organizations and volunteering at the Centro Cultural de la Raza in Balboa Park, by listening to a Spanish-language radio station, and by watching Netflix series and reading popular books portraying Latinxs in Southern California. In addition, Catherine followed the posts of several public Chicanx Organizations' Facebook groups to keep up to date on local events and discourse. She also visited relevant museums and attended academic workshops and events in the area.

In the original research design, we wanted to analyze the vigilance of those who could be perceived as migrants because of their phenotype or name and therefore face discrimination. The central research question was: How do these phenotypical "others" respond to intersecting state surveillance and civic observation?

On a taxi ride to the New Americans Museum in San Diego during her very first visit in November 2019, the Puerto Rican-Israeli driver wondered out loud what "New Americans" might be. He suggested that it might refer to the Anglo settlers of California. In fact, the museum seeks to educate about recent immigration to San Diego, with a particular emphasis on Mexican Americans. This episode suggests that which kind of San Diegans are perceived as "migrants" depends on whom you ask.

Reminiscent of the slogan "we are a people without borders," many Latinx U.S. citizens have mixed-status families and thus have positive or neutral attitudes towards migrants and may partially or fully identify as migrants themselves. It is also fairly common to identify with some kinds of migrants and not with others, based on immigration status, nationality, or race. For instance, a Costa Rican American man, who identifies as a "good migrant" because he pays taxes and helps his neighbors, distinguished himself from unauthorized migrants, whom he framed as a threat to U.S. national security. In addition, like Catherine herself, many Latinxs emphasize commonalities or differences with migrants (whether or not they identify as such) situationally, depending on the conversational context. For example, in multiple situations, Puerto Ricans expressed an ambivalent sense of belonging, sometimes foregrounding their status as U.S. citizens, other times referring to themselves as "immigrants." Both a Mexican American and a Honduran American interviewee expressed fear of potentially losing civic rights in the future, if anti-immigrant sentiment became stronger, even though they were both U.S. citizens by birth. Thus, who is a "migrant" not only depends on whom you ask, but also on when and how you ask.

Despite these ambivalences, there are few situations in which Latinxs who are U.S. citizens are genuinely mistaken for immigrants. In California, such situations include Border Patrol officers stopping cars driven by Latinxs on the assumption that they might be immigrants. Particularly in the context of border crossing, border guards do often make wrong assumptions based on phenotype and racial profiling. As Aura, a tough, strong-built *fronteriza* (U.S.-Mexico borderland woman) in her mid-thirties reported, she was once stopped by border guards when she was crossing the border at San Ysidro with her petite, younger sister of lighter phenotype and was accused of trafficking her. A Chicano San Ysidro resident similarly reported frequently having been asked for his papers at traffic stops. A "part-Mexican" woman said that on her family's trips to the desert bordering Arizona, her daughter was always the one to attract CBP attention because she had the darkest phenotype in the family.

Those are examples of racial profiling and there is not much one can do against one's phenotype. Thus, we found that racialized people in San Diego are cautious, but do not seem to employ strategies to actively avoid being mistaken for migrants. Indeed, in an opposite case, an Unión del Barrio member was often mistaken for ethnically Mexican American because he had a Chicano-style goatee and a bald head, as well as a Spanish surname but was in fact born to white parents and adopted by a Chicano family. Instead of seeking to leverage his birth whiteness, he chose a Chicano life.

Conservative-leaning interlocutors would employ strategies to differentiate themselves from migrants, such as by highlighting their U.S. citizenship and befriending or dating middle-class white people, earning some the joking slur of "coconut": Brown on the outside, white on the inside. This suggests that even if you are considered to "look Latinx" if you do not behave in stereotypical ways, you are not considered a "real Latinx." Thus, instead of seeking to blend in with Anglo-society entirely, many combined partial assimilation with openly expressing pride in certain aspects of their Latinx heritage, such as their national cuisine. In addition, some reported having been wrongly treated as migrants in situations in which they were with people of darker complexions or doing work associated with migrants, such as cleaning services. For many, blending into the dominant Anglo society was not an option available to them: "as a U.S. citizen, [they do] not feel politically represented or culturally included in the United States."[22] Thus, many of our interlocutors constantly anticipated potential discrimination and consciously or subconsciously planned for how to defend themselves once it did happen. There are certain wealthy, largely white parts of San Diego where Latinxs might be mistaken for migrants if they look poor, Indigenous or Mediterranean and/or speak Spanish too well. Yet Catherine observed that, because of how bilingual and multi-ethnic California is, Anglos rarely ask U.S.-born Latinxs where they

---

22 Galaviz, *Expressions of Membership and Belonging*, p. 51.

are from. At their most offensive, they might ask, "what are you?" as a way of eliciting someone's ancestry. In U.S. anti-immigrant rhetoric, there are usually less concerns about immigration in general and more about potential criminals and unauthorized people crossing the border illegally. Thus, when people are labelled as "migrants" by racist Anglo-Americans because they do not look Anglo-American, this "mistake" expresses a normative stance: the white supremacist view that the country belongs to Anglo Americans. Vigilance, then, is mostly about anticipating this type of racist discrimination.

Our study has focused on what kinds of watchfulness people adopt in response to being racialized as migrants and to experiencing racist discrimination. However, this did not solve the conundrum of how to find our research sample of migrantized people without ourselves enacting the racist gaze that marks them in this way. Nor did we wish to reinforce the coloniality that separates U.S. citizens from migrants within racialized communities. Further, as many anthropologists have shown, identity markers are not homogenous, essentialist categories, but are instead socially constructed, heterogeneous, dynamic, and ambiguous sets of intersecting relationships, which have concrete, observable meanings and effects in specific situations.[23] In order to avoid re-racializing people marked as "Latinxs" and "migrants," and in awareness of the problematic connotations of these concepts, we did not seek to eliminate ambivalences by working with precise definitions, but instead looked at inconsistencies in their usage as part of our analysis. Following Attia, Keskinkılıç, and Okcu, our aim was to critically deconstruct the migrant slot, not to impose hegemonic notions of racialized order through our research.[24] Accordingly, we centered our interlocutors' self-identifications and fluid positioning. Much like the mirror in the vignette at the beginning of this chapter, we focused our research on contexts that are historically associated with Chicanx and Latinx communities in San Diego without assuming that all people that stepped in front of our research lens could be defined as "Chicanxs" and "Latinxs." The Centro Cultural de la Raza, despite being created by and for Mexican Americans and Latinxs, was in fact a highly inclusive space that was welcoming to all respectful visitors and volunteers, regardless of their ethnoracial and national identity or immigration status. Similarly, other Chicanx organizations were open to accepting trustworthy members of other ethnic backgrounds, which is also what made these spaces accessible to our team.

---

23 Attia/Keskinkılıç/Okcu, *Muslimischsein im Sicherheitsdiskurs*, p. 52.
24 Ibid., p. 55.

The Border as mirror multiplied

Among European anthropologists, walking has long been considered an important way of producing knowledge. According to Tim Ingold, "it is by walking along from place to place, and not by building up from local particulars, that we come to know what we do."[25] He argues that, rather than being transmitted or made, knowledge grows from embodied improvisations when engaging with the world as an open-ended process. In this spirit, Catherine spent a large portion of her 10-day first visit to San Diego in November 2019 walking through different parts of San Diego, and on their arrival in San Diego in September 2021, getting to know the city by walking through its neighborhoods was also something that Eveline, Carolin and Jonathan were initially keen to do.

During Catherine's first ten days in San Diego she also walked across the border at San Ysidro to visit Tijuana with a tour guide, Rogelio, a Tijuanense in his early 20s. While passing through security to enter Tijuana at lunch time was a matter of minutes, crossing back to San Diego at sunset took approximately an hour. Gesturing at the long line of mostly Mexican-looking people, Rogelio said that this is one of the key crossing times, as many people prefer to cross on Sunday night if they have to work early on Monday morning. The heavily armed border guards appeared highly vigilant, making sure that the line was orderly, and that nobody skipped ahead – although one older lady managed to do so anyway. Catherine felt scrutinized, noting that the checks on this side were generally more rigorous and more intimidating, although she passed without problem. Rogelio said that he was not concerned about experiencing discrimination in the U.S., as you can experience discrimination in Mexico as well. However, he stayed *trucha* in Tijuana, which is to say, he was much more cautious there than in the States because if someone stole his passport and did something illegal in his name in the U.S., he would permanently lose the ability to live a middle-class lifestyle, which depends on being able to pass the border with ease to work in San Diego.[26] He mentioned that there used to be no physical border between the cities, so that it used to be possible to go partying on the beach and then, drunkenly stroll past the police officers with beer in hand to get to the other side without fear.

Making her way to Barrio Logan from the harbor via Seaport Village and Imperial Street on another day, Catherine realized that to be a *flâneur* in the style of Walter Benjamin was far more challenging in Southern California than in Tijuana,

---

25 Ingold, Footprints Through the Weather World, p. 121 f.
26 Cf. Yeh, *Passing*.

or in the European contexts she was familiar with.[27] The 12th Street/Imperial trolley station is a place where many unhoused people stay. A man was screaming at some imagined adversary. Catherine tried to avoid walking too close to anyone but felt conspicuously middle-class and white. As Barrio Logan was coming into view, she walked past two men. One quipped tauntingly, "Nice shoes!" The other, "Nice sunglasses!" She pressed on, heartbeat quickening, feeling uncomfortable. As a walker, she stood out because in San Diego, middle-class people only walk in specific, "nice" parts of town, where there are parks, shops, and eateries. It is highly unusual for them to walk long distances, unless they are attempting an "urban hike" to specific destinations, such as the various bridges of Downtown San Diego. Jonathan felt similarly when later walking alone from the 12th Street/Imperial trolley station towards Barrio Logan, a route which on foot involves walking for several minutes past tents and unhoused people on the street. Jonathan tended to take the bus into downtown San Diego, which often involved changing buses at City College in downtown, where unhoused people often sit close by with shopping trolleys of their belongings, which can sometimes feel disturbing, as some individuals behave erratically.

Generally, San Diegans rely on their cars to get around town, even when they only have to go a short distance, as pedestrian paths are often awkward or even missing altogether, particularly in the areas between different neighborhoods. Multi-lane expressways and bad sidewalks characterize the cityscape. Thus, driving multiple times a day is common sense for most San Diegans who live outside of the more walkable Downtown area. Accordingly, some teenagers from the urban center of San Diego have never seen the ocean, even though it is only half an hour away by car, as one former volunteer at an organization facilitating ocean visits for such deprived youths explained. When using the public transportation system himself during the team fieldwork phase, Jonathan was able to confirm many of Catherine's impressions and found that trolley and bus riders as opposed to car drivers were segregated groups.

In summary, even though walking proved challenging in San Diego, having to negotiate dangerous crossings and feeling anxious when passing through the highly militarized U.S.-Mexico border, Catherine was able to gain an embodied knowledge of the materially segregated nature of the city, the way in which various physical blockages limit the movement of the poor across national and neighborhood borders, and the high importance of material wealth for San Diegans in the absence of a strong social welfare system. The social anxieties that are tied to these factors are key motivators of watchfulness in San Diego.

---

27 Benjamin, *Stadt des Flaneurs*.

Occasional taxi trips at the beginning of fieldwork were expensive but informative as well, as drivers were often happy to talk about their lives and racialized perceptions of threat. For example, a 24-year-old, cheerful, white-read and tattooed Asian San Diegan Lyft driver said that he had not thought about vigilance before, as he did not feel particularly unsafe anywhere. However, he did not like working in Downtown San Diego at night anymore. He claimed that there are too many bar fights, drunks and meth addicts there, and instead recommended going to Tijuana, which he described in a stereotypical way: "TJ's a crazy place [...] you go there for cheap Tequila and hookers." He also warned Catherine to "watch out for pickpockets," because a friend of his had his passport stolen and was stranded there for a bit, as the U.S. are very strict about requiring full documentation. He also warned against wearing expensive jewelry in Tijuana, even if it is fake, as people might rip earrings straight off your ears. "And watch out for corrupt cops, they'll look for ways to make you bribe them." On another trip, a mustachioed Lyft driver, speaking with a Mexican accent, told Catherine that he used to live in National City, claiming that "it was bad there." People had been very afraid of crime, so they often had iron bars in front of their windows, which is also the case for many homes in Barrio Logan. Catherine was surprised when the driver added, "there are too many Blacks and Mexicans living there," implying that he did not identify as Mexican, or at least distancing himself from other Mexicans. His daughter's bag was stolen by a Black kid in primary school, he explained. So his family moved to the north, to Mira Mesa, where he maintained that they can live "more peacefully." This reminded Catherine of Lara, an Asian student's comment that her parents had warned her to stay away from other minorities at her "Hispanic-serving" university, San Diego State University (SDSU): "Don't let them drag you down." Yet this warning had only made the young woman more curious about other groups, so that she did not avoid anyone. However, Lara did consider the area surrounding SDSU to be unsafe. There had been a shoot-out in front of her dorm. Winter term was particularly stressful for her because she tried to get home at 5pm, before dark. As a result, she and other students felt restricted in their movements, mostly sticking to two streets near the campus to find lunch.

Overall, walking and taxi rides along with conversations with San Diegans showed that San Diego is a border city in more than one sense. As disadvantaged people face many barriers to movement in San Diego, both in terms of the physical infrastructure as well as fears about their safety, the borders that act as mirrors to their identity are multiple.

Room hunting in a landscape of anxiety

As Joost Fontein observed, "There was a (perhaps mythical) time in anthropology when living with informants in the field was perceived as an essential part of ethnographic fieldwork."[28] He suggests instead considering practical factors, such as needing a desk or transport, and how one's presence will affect the people one is working with, regardless of where the researcher chooses to live.[29] With this in mind, Catherine attempted to find a host family in Barrio Logan and first rented an Airbnb room in nearby Sherman Heights. However, she found that the very watchfulness that she was interested in studying immersively became an impediment to finding a base in Barrio Logan.

The people there relied on watchfulness among neighbors to keep their homes safe, as well as on their dogs. The Santa Ana-based Mexican American journalist Gustavo Arellano explained that this is the case in Latinx neighborhoods throughout Southern California: "All barrios have these dogs bark, usually, not just a Pitbull, it's like the extreme, either Pitbull, Boxer, or Chihuahua. You need barkers to immediately let you know like hey, someone's at the door. Who needs a Ring [high-tech doorbell] when you have a dog, you know?" He added, "Dogs can be ferocious, at the very least they're hypervigilant. They're always out there, they distrust people that they don't know." As there were hardly any options available, given that most families in Barrio Logan had barely enough space for themselves and thus were not renting out rooms, and she did not want to live in a household with aggressive dogs, Catherine was not able to find a host family in Barrio Logan. She had also been warned against doing so by white Anglo potential host moms:

> I would suggest not taking a home stay in Barrio Logan, Logan Heights, Golden Hill, Sherman Heights, etc. Those neighborhoods are lower socioeconomic neighborhoods that are slowly transitioning due to their proximity to downtown San Diego, but they are just not safe. Barrio Logan has some lovely old buildings and really interesting businesses opening up and housing there will be less expensive than other parts of town, but risky.

Therefore, Catherine eventually accepted a room in what was presented as a comparatively safer, quiet, middle-class suburban Clairemont Olive Grove neighborhood in the northern part of San Diego, where signs featuring suspicious eyes warn thieves of a Neighborhood Watch – relics of the 90s, as the watch no longer exists. Instead, many people in Clairemont and similar neighborhoods have installed Ring security systems, which are operated by Amazon. The police can access

---

28 Fontein, Doing research, p. 71.
29 Ibid.

Ring camera footage of suspicious activities. Moreover, many San Diegans use the app Nextdoor to warn their neighbors of suspicious activities and photographic evidence, including from their Ring doorbells. As Lara explained, this technology is likely to be used more in white areas, where there is generally more tech and Wi-Fi available, whereas in a place like Barrio Logan, people rely more on knowing each other and text safety warnings to each other directly. It took several months before Catherine was able to access the app, as one of her neighbors had to confirm her identity in order to grant her admission. Even more popular than sunsets and pets, safety-related content makes up approximately half of the top posts, with examples including (spelling as in original):

> "Brazen car break in at 645am. Woke up this morning to this guy breaking into a car. Redneck sticker on the windshield and the car is registered out of Chula […]."
>
> "Sketchy guy – So sitting in my car and this winner walks rite in front of me looking dead into my car […]."
>
> "Stolen Property – BEWARE: Brazen thieves on January 6, 2022 x 1:15am stole our Catalytic Converter. There were a lot lights on […]."
>
> "Man trying to open front door. – Friendly reminder to keep your doors locked. This man tried opening our front door a few minutes ago and our car […]."
>
> "Hit and Run. – On 1/2/21 at 4:15 pm I was turning left from prospect to fay and my Mercedes gla was hit on the right passenger side by a red […]."
>
> "Thieves stealing packages. – These people are stealing packages. They stole one from me and I tracked them Down based off of the ring video. They are in bay park […]."
>
> "Stay inside & lock your doors – Denver / Clairemont Dr / morena area stabbing a few minutes ago. Police Looking for suspect Hispanic male with tattoos […]."

The San Diego Police Department also sends out warnings via the app. It is not difficult to see how the app fosters not only observation but also a sense of paranoia. The steady flow of warning messages about, in many cases, unconfirmed suspicions about potential crimes can easily give the impression that crime is on the rise, whether or not that actually matches official statistics. Pre-existing classist and racist perception biases, such as associations of "rednecks" and "Hispanics" with danger, can find a large audience instantly and are amplified, thereby reproducing, and strengthening, stereotypical views about certain, often disadvantaged populations.[30] While Catherine did not systematically analyze Nextdoor posts, lurking on the app gave her insight into the landscape of anxiety that many of her neighbors lived in.

---

30 Makena, Inside Nextdoor's 'Karen Problem'.

In summary, both in Barrio Logan and in Clairemont, attention was not randomly directed at anyone, but followed racialized and classed patterns. The living conditions and preferred technologies that people used to assist in their alertness differed across these socioeconomically and ethnically distinct neighborhoods, which meant that rooms were far more accessible to Catherine in Clairemont than in Barrio Logan. Yet far more critical for the success of our project were the difficulties our team experienced with respect to research access and safety related to the wider global situation: the lockdown conditions that were imposed in response to the COVID-19 pandemic.

The upheavals of COVID-19

When she began fieldwork in February 2020, Catherine approached ethnographic research in a fairly conventional way, such as by walking through neighborhoods of interest, visiting relevant museums, meeting with local researchers, and networking at community events. Among others, she attended meetings of Chicanx organizations, such as Unión del Barrio, where she was invited to volunteer with their community patrol. She also signed up for the Chicano Park Day trash crew. Participant observation, "the quintessential ethnographic fieldwork method,"[31] is an embodied, multisensory, and somewhat experimental practice that involves "learning through doing and experiencing as much as through watching and listening."[32] Thus, ethnographic research usually entails deep immersion in the context studied, often relying on a large number of in-person interactions. In addition, in regular meetings, Catherine discussed her observations and experiences with Eveline on a regular basis. Thus, they reflected together on how the fieldwork evolved. This was the more important as Eveline, who had designed the research as Principal Investigator and managed the project throughout, had planned to conduct a joint fieldwork together with Catherine as a core element of this project.

Yet when the pandemic lockdown started, these plans that had been made came undone and there was no way Eveline could travel to San Diego. Beyond the lockdown guidelines of the U.S. government, an official email from Catherine's host institution, the University of California San Diego, banned in-person fieldwork from March 17, 2020, onwards. Catherine had to cancel volunteering with the community patrol as well as the border tour she had been arranging with Customs and Border Patrol via the sociologist Irene Vega at the University of California Irvine.

---

[31] Fontein, Doing research, p. 75.
[32] Ibid.

The frustrating and uncertain situation made it tempting to fly home. However, Catherine, in consultation with Eveline, decided to stay in San Diego and adapt the research plan instead.

While Catherine's initial focus was on the Chicanx neighborhood of Barrio Logan as a field site, the research focus shifted, as public health measures limited movement within the city, and public gatherings were banned. Thus, first-hand observations were only possible within Catherine's home neighborhood, Clairemont, and in supermarkets, on beaches, and parks. Our research methods thus temporarily changed from participant observation and face-to-face interviews to using digital ethnographic[33] and auto-ethnographic methods.[34] Some interviews were conducted online, and more weight had to be given to the use of secondary sources, particularly social media (Instagram and Facebook). Catherine also continued to take part in the Centro Cultural de la Raza's *círculo de mujeres*/Women's Circle, which had shifted to meeting on Google Hangouts, and she also attended a Barrio Logan-based Open Mic night, La Palabra, via Zoom. The brief phase of initial participant observation helped to provide greater context with which to analyze the data gathered online. Catherine was also able to maintain the friendships she had already made via text messages, calls, and WhatsApp. While her circle of interactions was now much more limited in size, the relationships became closer and more complex.

Similarly, digital participant observation became central to Carolin's research project, which she conducted between March and July 2021 for a total of four weeks for her master thesis. She started with her research partners' Instagram account, which she understood as a place to which their life extended and thus made itself visible in a unique way. She also paid attention to how Instagram was used in specific ways in this cyber-social network to establish solidarity references, create awareness and new meanings, and share healing methods.[35] Carolin thus carried this participant observation out by noting the activities of the group via Instagram – both that of the *brujxs* (witches) as the central group of her research themselves and that of other activists, customers, and other followers within their network (more in chapter 6). To identify discursive, affective, and symbolic meaning, she used screenshots and screen recording and took notes.

When Carolin was later able to do field research with the rest of the team for two weeks in San Diego in September and October 2021, this gave her deeper insight into the context in which her research partners lived and engaged with. In

---

33 Pink et al., *Digital Ethnography*.
34 Behar, *The Vulnerable Observer*.
35 Hine, *Ethnography for the Internet*, pp. 22 ff.

addition, she was able to meet with the central group of *brujxs* that she had previously only been able to talk to via Zoom. In relation to social inequality and social media activism, Carolin was able to refine her understanding of certain meanings and narratives evoked on Instagram through digital and on-site interviews.[36] This led to new insights and helped to build rapport and trust, a point we will return to later.

Catherine was particularly grateful to live in a "research-related" family, as her host mother Christina referred to herself as "Mexican" (her father's nation of origin), enabling an auto-ethnographic approach. Over dinner in February 2020, her white Anglo-American host father Pete asked Catherine what people in Europe thought about Trump, keen to hear her personal opinion. Catherine was open about disliking the current president. Her hosts then revealed themselves to be ardent Evangelical Republicans who had voted for Trump because they believed that he defended the U.S. as a "Christian country," while Obama had said that "the U.S. is no longer (only) a Christian country." They were also against abortion, transgenderism, and socialism, and for capitalism and strong borders that they believed would keep out Mexican cartels.

Catherine's hosts took an interest in her interpretations of the sociopolitical climate in San Diego. Similarly, Catherine was interested in their perspective, as a way of balancing the often radically left, anti-establishment and decolonial views and agendas of her Chicanx interlocutors with an opposed conservative viewpoint. During the first two months of living there, Catherine additionally benefitted from conversations with one of the fellow tenants who was a police officer in training, which led to informative conversations about racial profiling and gun control, among other topics. While her hosts were kind, upbeat, and very generous in introducing Catherine to their family and friends, their political disagreements and occasional anxieties surrounding anti-COVID hygiene meant that Catherine could not fully relax at home, as there was a sense of constantly mutually watching and judging each other.

For instance, in October 2020, over breakfast, Christina shared that her friend had waved at a Trump motorcade only to get shouted at with profanity by another bystander. Catherine pointed out that she had observed similar behavior from Trump supporters against BLM protesters. This seemed to upset Christina, because she emphasized that she had been sharing about something bad that had happened to her friend. She had wanted an empathetic reaction, rather than a blame-splitting one.

---

36 See Barassi, Social Media Activism, p. 410.

However, there were phases of respite from the pandemic lockdown conditions. When protests erupted in response to the killing of George Floyd, UCSD communicated that university members were not banned from participating in Black Lives Matter protests. As she cared about this cause, Catherine accompanied Latinx and other POC protesters marching in solidarity, which later became the subject of several of our research publications.[37]

Eventually, the COVID situation improved somewhat, and Catherine took the opportunity to volunteer outdoors at the Centro Cultural de la Raza on Saturdays. Catherine took part in the Centro garden transformation, whenever she had a chance. However, after missing a few gardening sessions, she was greeted with "long time no see" – a reminder that, *tequio* (collective work), as it is understood by the Centro, requires ongoing, dutiful, shared work. This leads us to the complexities of rapport during fieldwork.

Relationship rupture and repair

In the Centro Cultural de la Raza's zine "Raza Visions II," the collective Xicanxpatista Autonomxs cite Cherrie L. Moraga as saying, "The real power, as you and I well know, is collective. I can't afford to be afraid of you, nor you of me. If it takes head-on collisions, let's do it: this polite timidity is killing us."

Generally, participant observation requires anthropologists to spend a significant amount of time with their research participants, getting to know key interlocutors intimately, as well as understanding the society in which they live, by engaging in their daily activities alongside them. To work with Chicanx organizations requires an even deeper level of immersion and commitment from researchers, as Chicanxs are highly alert to the possibility of research relationships becoming exploitative means to simply extract data. As one Mexican American board member of the Centro Cultural told Catherine, a relationship where both sides benefit from the exchange is acceptable but "everything would be easier if people didn't need money." Generally, Chicanxs demand of researchers to become accomplices, rather than mere allies in principle, for their struggle. Commitment cannot simply be stated but needs to be proven with repeated actions over time.

Thus, trust needs to be built long-term and is vulnerable to being lost. According to Danny Jorgensen, "The quality of data is improved when the participant observer establishes and sustains trusting and cooperative relationships with the

---

37 Whittaker/Dürr, Vigilance, Knowledge, and De/Colonization; Dürr/Whittaker, "Go back to your country!"

people in the field."[38] However, Sökefeld and Strasser argue that "the mantra of building trust with our research participants becomes questionable during ethnographic fieldwork that is pervaded with tension, contradictions and anxiety whilst under surveillance."[39] We would like to argue that, in the highly surveilled border city of San Diego, where Chicanxs are raised to be cautious, trusting relationships with research participants are in fact very important, but rarely evolve in a linear way. If trust is rebuilt after a moment of conflict because the parties involved care enough about each other to do so, relationships can even become closer after reconciliation. Following Moraga, many Chicanxs do not seek to avoid head-on collisions. Collective struggle also demands confrontations between those who are struggling together that arise from holding each other accountable. Accordingly, the relationship between trust and conflict is complex in this context. Catherine thus experienced her relationships with fieldwork participants as a constant up and down.

Throughout her fieldwork, attending the Centro Cultural's bi-monthly women's circle had been one of the few reliable constants for Catherine, like an emotional anchor. According to its Puerto Rican organizer, the purpose of the circle was to provide a safe space for collectively processing the trauma of living under lockdown conditions. As a description of the circle on Facebook stated:

> We see how our communities are uniting in ways beyond the physical, reaching out and offering presence and support. It's a profound act of kindness to make space for connection through a sort of solitude, it proves more and more that our strength as individuals creates the collective woven network that when needed springs into vibration like a silken web. Soft, strong, vital, and under all circumstances repairable and adaptable. This is the common goal of the Women's Circle, to make space and share time with our community with the intention of support, aimed at healing and learning.

The first time Catherine attended the circle was in November 2019. The Centro Cultural building has a round, flat shape and is covered, inside and outside, in striking, colorful murals, which take inspiration from Mesoamerican lore. On the registration desk lay an array of information leaflets directed at unauthorized people and those who might want to help them, including legal advice. Then Catherine shyly walked over to a 29-year-old woman with long dark brown hair and an intricate arm tattoo peeking through her loose-fitting top, who was arranging seats in a circle around a round blanket. It had a candle in the middle, surrounded by some sheets of paper, which were reminders of the code of conduct. Catherine

---

38 Jorgensen, *Participant Observation*, p. 69.
39 Sökefeld/Strasser, Under suspicious eyes, p. 161.

introduced herself in Spanish, asking if she was at the right place. The woman, Berenice, said yes and suggested that they could speak English, unless Catherine wanted to practice her Spanish, adding that the circle would be in English. Catherine told her that practicing is good, but that she considered it a matter of respect not to privilege English over Spanish, which made Berenice smile. She said Catherine was welcome to take part in the circle and asked if she had ever been to anything similar before, which she had (in Mexico). Two more women arrived, who already knew Berenice and had taken part in the circle before. They began chatting, and then Berenice offered tea and invited all to make themselves comfortable. After the circle already started, a third woman joined them. Berenice explained the rules, emphasizing that this was a safe space where participants were only supposed to make first-person statements and needed to be respectful and avoid making accusations. Sessions typically began with a meditation and a check in regarding participants' emotional states. Then Berenice would introduce a guiding question for the group and participants were given opportunity to share their views and experiences with it. Guiding questions often touched on difficult topics, such as conflict and decolonization. The sessions were strictly confidential, so that women and non-binary participants could express themselves freely. This meant that they could not write about anything that was said during the sessions, but hearing about attendees' diverse hopes and struggles was still very helpful for understanding their lifeworlds better. In addition, Berenice became a friend of Catherine's and agreed to be interviewed for the project, as she found the topic important. A PhD student (acupuncture and alternative medicine) herself, she could relate to Catherine as a fellow researcher.

Towards the end of March 2020, Catherine suggested that it could be a nice idea for the group to write something shareable for the wider community. The other women discussed this among themselves and then expressed worry about "the ownership of the material," "who would be able to access the document," and the group as a "safe space of confidentiality," highlighting that they did not want to do anything that would change the perception of the space and make people feel less safe or obligated to share. Catherine explained that this was a misunderstanding. She was not looking publish anything academically at this point yet. The idea had been to collect some basic thoughts surrounding the pandemic situation, without any sensitive information attached to it. This had in fact been inspired by what one of the group members had said about how as a group they could function like a kind of "contagion," multiplying kindness through their personal networks. Catherine's hope had been to create bite-size content that could be shared on social media to reach people who were not joining the circle for whatever reason but may nonetheless have been interested in some way, or would have benefited from knowing that they were not alone, that others shared their grief and other feelings, that there were good sources of guidance beyond those of the government, and that there were venues for support and resources. Catherine also said that it is fine if nothing comes of this. They ended by thanking each other. Catherine was left feeling deflated, like she would always be perceived as some kind of opportunistic leech. She had believed that this was not the case in the *círculo* but had to realize her mistake. However, this episode did not change her participation in the circle, and it did not appear like any serious damage had occurred.

Yet there were cases in which research relationships ended without recovery. In one case, Catherine distanced herself from a Mexicano man in his early thirties who had taken a romantic interest in her. The most impactful falling out happened with Catherine's host family. Due to their ideological differences, their relationship had always been tense, despite being relatively close, even affectionate. This tension came to a head on election night. Pete provocatively asked whom she would have voted for. Catherine attempted to avoid conflict by suggesting that, as a non-voter, it would not be her place to say. He then quipped that "it doesn't affect you anyway." Catherine responded that this was incorrect, as the parties' differences in migration politics did personally affect her. This led to Christina angrily pointing out that it was Trump's government that had granted her research visa, which started an explosive argument. Catherine moved out a few days later. Her new host was an Iraq-born American Civil Liberties Union activist, whom she interviewed about vigilance and racial profiling for our project. In her new home in the ethnically diverse, upper middle-class Midtown area near the San Diego airport, Catherine felt more respected and relaxed. Meanwhile, Catherine and her former hosts expressed words of peace to each other through a mutual friend, Libertad, a young Dominican who had also briefly rented a room there, but they have not met in person again.

Rupture and repair also occurred in our research team, as Catherine changed employers after the first 18 months of the project, meaning that the work plan had to be substantially revised, and the team grew because Jonathan replaced her as a postdoctoral researcher. It also meant that group fieldwork became an even more important, if brief, phase in the project. Yet before expanding on this topic, let us discuss the implications of rupture and repair dynamics for our positionality as researchers through the example of Carolin's fieldwork experiences, who acted as research assistant in the project while also conducting research for her master's thesis digitally and during our joint research stay in San Diego, as mentioned above.

Positionality, intersectionality, and emerging research subjectivities

During all aspects of Carolin's research, she was confronted and confronted herself with positionality as an important factor for access to and relations with research partners and an essential one for interpreting any results. She defines positionality as the position of a person within existing power and hierarchical structures, which decisively shapes the living environment as well as any research. Reflecting on and making one's own positionality transparent is particularly important in anthropological research in order to make it explicit that knowledge production is socially situated and not independent of one's own living environment.[40] Race and white supremacy not only have formative consequences for the positioning and identity of research partners but have also been identified as important for the anthropological approach and associated ethical questions.[41] Carolin started from self-descriptions and denounced attributions from others and understood the overlapping and multiple marginalization in an intersectional way.[42] Drawing on Rocío R. García, she refused to view identities as variables and disadvantages homogeneously but strove for a relational approach that places sociocultural processes at the center and depicts the variation and fluidity of interaction arrangements and social relationships.[43] In addition, García always understands these positionalities in relation to struggles for social justice, which were also at the center of her own research.[44] Carolin understood intersectionality on the one hand as the visualization of dynamics between her and her research partners and their experiences of discrimination, but it also shaped her theoretical and methodological approach.[45] It was important to her not to regard these notions of differences as essentializing and simplistic, but as a starting point for the analysis of formative and real power relations.[46]

The experience and lives of the *brujxs* she was working with, and their network was shaped by various experiences of discrimination that they made as a non-binary, queer Black, Indigenous and People of Color with family migration experiences and multiple intersections of oppression. Carolin's life, however, was

---

40 Speed, Crossroads of Human Rights, p. 74.
41 See Rosa/Bonilla, Deprovincializing Trump; Beliso-De Jesús/Pierre, Anthropology of White Supremacy; Rosa/Díaz, Raciontologies.
42 Chavez-Dueñas et al., Healing Ethno-Racial Trauma.
43 García, Politics of Erased Migrations, p. 4f.
44 Ibid.
45 Cho/Crenshaw/McCall, Field of Intersectionality Studies, p. 785.
46 Anthias, Translocational Positionality.

shaped by many privileges, which led to constant negotiations of positions and the confrontation with prejudices and oppressive elements. Especially during the interviews, her research partners made clear that they distrusted her as being part of problematic systems. She was made to actively question the partially neocolonial nature of anthropologists' relationships with their research partners.[47] Consequently, Carolin's research partners openly distrusted her as a white privileged woman interested in their spiritual practices. They also distrusted her as a researcher and the university institution that she represented. This distrust stemmed from the fear of their healing techniques becoming "mainstream" and for them to lose their integrity in openly talking to her. However, there was also the concern that privileged individuals could co-opt their medicine and thus capitalize off of it. This could result in the people who needed it most, and from whose ancestors the techniques originally came, no longer being able to afford their own medicine. During these conversations it became clear to Carolin that an explicit reflection of herself as a researcher and of the institution university was required as part of her research.

Carolin often found herself in the paradoxical situation of having to fully agree with her research partners in their distrust and justified concerns and at the same time having to be convinced of her research. She recognized the potential exoticization of research partners as a serious problem in anthropology. Thus, she tried to explain that her research was an attempt to acknowledge problematic structures and to decolonize her own research process, while doubting that this was within her capabilities. In addition, Carolin had difficulties with working with 'fixed' and inadequate categorizations. These can be seen as essentializing and not going far enough in descriptions of identity and subjectivation. However, the descriptions she was using also came from the field itself, since they were intended to depict the consequences of real power structures. She simultaneously recognized the fluidity and flexibility of these categorizations.

After establishing some basic trust, the *brujxs* told Carolin what expectations they had of her and her work. Their hope was that the cooperation would result in a work that could inspire other people to get involved in radicalism and a "revolution" in their sense. Because of Carolin's local and personal disconnection from the field and struggles, she felt it necessary to hold herself accountable, connect beyond research, and form an alliance: she shared the political goal of fighting social injustice.[48] Despite all the hurdles she encountered in the course of her research – both in terms of the pandemic and the relationship with her research

---

47 See Speed, Crossroads of Human Rights, p. 70.
48 Speed, Crossroads of Human Rights, p. 72.

partners – her interactions with the *brujxs* were very fruitful and valuable for her findings. Carolin tried to recognize and incorporate the (initial) mistrust and expectations of her as a researcher as constitutive of the process.

Group fieldwork

Part of the data collection and the writing process were collaborative between the project team. A joint field research in San Diego conducted by all team members took place in September and October 2021. The main goal of our joint research was to make a transition after Catherine's departure from the project that would allow the project to continue and to equip us with a common ground for our further discussions. Thus, this joint field research stay was essential in order to manage the collaborative interpretation and reflection of the material and thus, to co-author this book.

In the first days, we visited together a range of socio-economically different neighborhoods, most importantly Logan Heights and Barrio Logan, but also La Jolla, Coronado Island, Lemon Grove, and Chula Vista, for example. We also spent time in parks and specific spaces, such as Chicano Park, Balboa Park, and the Border field state park, and visited diverse museums to better understand the representational politics of the city and its powerful geo-political and economic forces, as displayed in the Maritime Museum of San Diego. We included also the Kumeyaay museum to acknowledge the ancestral homeland of the Kumeyaay Nation and to extend our respect to Native Californians. These visits helped us in developing an understanding of the multiple perspectives on the city's history and how historical trajectories shape its uneven racialized social fabric in the present. We were also able to spontaneously take in a pow-wow of local First Nations peoples in Balboa Park, which made us even more aware of the vibrancy of Native American life, but also of Indigenous resistance, calling out that "you are on stolen land."

During these first days together, we did not talk to many other people yet, but more amongst ourselves, and rather soaked in the urban environment. However, this turned out to be a rather different experience for each one of us. While Eveline noted how much her observations of the urban environment were shaped by her prior readings on San Diego, Jonathan's perception was less cognitive but more affective. He physically sensed the pollution in certain areas of the city strongly and felt uncomfortable, in particular beneath the Coronado Bridge in Barrio Logan and thus experienced concerns about pollution by residents of Barrio Logan himself. Jonathan thus had to constantly clear his nose and wear a mask. To this day, environmental injustices are pertinent in San Diego and affect urban dwellers' lives and well-being. Carolin, meanwhile, tried to absorb as many facets as possible of San Diego, because her own research project formed the basis for her master's thesis, which depended on generating ethnographic material during this stay. This led her to meticulously document as many aspects as possible during the fieldwork, in fear of missing or forgetting something.

During their time together in San Diego, Catherine became a gatekeeper for the team, acting as a guide and group leader. As she knew the city well, the team benefited from her knowledge. This was significant because the research process involved gathering data through establishing and maintaining social relationships – rather than being confined to observations and reflection amongst ourselves. By taking on the role of someone with local knowledge, Catherine became sensitized to the power imbalances involved in research relationships and worrying about the ways in which her comments might be (mis)understood and used by others. Even though we were all (white) anthropologists, there were different fieldwork experiences and academic positions within the group, and these emerged in social interactions with our research partners. For instance, the other group members noted Catherine's fear of us transgressing social protocols thereby forfeiting trust that she had painstakingly built over many months. For example, she would point out when something was framed from a white perspective. This also helped Catherine notice just how much she had learned to practice alertness in a routinized way. She realized that she had learned to cautiously mitigate her whiteness in an anticipatory mode, lest she or the project be perceived as being racist and/or colonialist. Beyond the issue of social protocols, adding additional team members to existing research relationships altered the dynamic between researcher and research subjects, which was particularly difficult to navigate in the case of Catherine's relationships that thrived on intimacy and confidentiality. For instance, Libertad cancelled a meeting after Catherine asked if she could bring someone else from the team along, too, in light of shared interests.

Similarly, the dynamics in the group changed over the course of the stay and we split up into different constellations or went alone. Carolin got in touch with research partners in the context of her master's thesis, while Jonathan and Eveline explored new avenues for a possible follow-up project. This provided Catherine with more time to maintain her contacts as usual – even if she wanted to know where we were and who we were talking to about what after our encounters. One afternoon, Jonathan and Eveline told Catherine that they had talked in the street of Barrio Logan to one particular person, who was denounced and called out as a "traitor" by some people. Catherine was worried that our conversation with him in public could damage her and our social relationships with some interlocutors in Barrio Logan. This concern can be read in multiple ways. Catherine's caution in the group and sense of care and responsibility towards her research contacts extended to the phases of the fieldwork when she was not physically present, as, much like the whole team, she was aware of the coloniality of research in the context of universities. However, this was also about sharing worries, doubts, and information – and thus co-producing both care and knowledge between us as a team. It is also important to note that while we may have seen ourselves as a group and felt to some extent responsible for each other's actions, our interlocutors may have well differentiated between us as individuals.

Other strains and tension surrounded the issue of coordination. On the one hand, this was facilitated by the fact that almost all of us lived together (Jonathan lived separately in other neighborhoods of San Diego, first in Rancho Peñasquitos, and then in an Airbnb in the Chicano neighborhood of National City with a community organizer), even if this close proximity was not always productive. For example, there are different needs for closeness, distance, and withdrawal, which could not always be guaranteed. Nonetheless, the process of conducting group fieldwork was productive, as it allowed us to observe not just our research subjects but also to observe how our research subjects responded to the other members of the research team. It also provided the opportunity to exchange our own observations with one another in situ, and build on these in our excursions, rather than just discussing these together after the fact. In addition, conducting research together protected one another. For example, Catherine had been concerned about whether she would be safe meeting with a certain male individual on her own. Interviewing him as a group made her security concerns disappear.

These multiple constellations of watchfulness as a methodological principle were formative for the entire research stay and we systematically integrated it into our reflections. Overall, working in a group multiplied the layers of reflexivity normally involved in ethnographic research.[49] We not only exchanged research-guided ideas about both our observations of specific situations and conversations, but also about our mutual observations of our own practices in the research process. The observation of ourselves as researchers added another layer of observation to our group research. For example, Jonathan observed Eveline and Carolin's behavior at the border and noted them photographing "the other side" through the fence. While for Eveline and Carolin it was more an act of attention and interaction, Jonathan read it more as a voyeuristic practice – though in the discussion that followed we noted that those on the Mexican side of the fence were watching us as much as we were watching them. However, this observation of observation

---

49 Turunen et al., Poly-Space.

made us reflect about our practices of "observing" as ethnographic method and led us to discussions about differentiating watchfulness from voyeurism.[50]

In the course of our joint field research stay, we all developed our own relationships with some of the interlocutors through our own telling of our biographies, concerns and perspectives. Intense conversations ensued over lunch at a Mexican restaurant, where we talked about racism and discrimination in San Diego in comparison to Germany. Aura described herself as a "fronteriza" and spoke at length about her borderland experiences. By acknowledging her situated knowledge,[51] our roles shifted in the group – and Aura transformed from a speaker into an expert, and we turned from listeners into learners. Aura's knowledge came to be valued through her emphatic narrative, and we showed that we valued her interpretation of the world through this arrangement. Through these conversations, an interactive framework emerged that fostered collaborative knowledge production as Aura wanted to know about racism in Germany as well. In this kind of exchange – by emphatic talking and storytelling about everyday life, but also about specific issues of concern for all in the group – lies a transformative potential with constantly changing roles that can ultimately lead to a new form of "we" in the research process.[52]

While we as a group shared these intense fieldwork experiences, we also had divergent experiences because of our diverse social backgrounds and tasks "at home," which can never be left entirely behind but rather stretch on into the field work context.[53] For instance, Eveline maintained her role as a professor throughout, which meant that she was still involved in online teaching while being physically in San Diego, lecturing across multiple time zones, and actively participating in conferences. This was similar for Catherine as well. These responsibilities continually drew attention to dimensions unrelated to fieldwork that complicated immersion in the field. Personal aspects also occurred that were particularly emotional for individual members of the group, such as the death of family members, which dispersed the attention usually given to the field context.

The writing process

The joint fieldwork combined with, and was necessary because of, the collaborative nature of the writing process. Since Catherine left LMU to join Goethe Univer-

---

50 See Dürr, Beobachter:in.
51 Haraway, Situated Knowledges.
52 Cf. Siry/Ali-Khan/Zuss, Cultures in the making; Dürr, Gemeinsames Beobachten.
53 Dürr/Sökefeld, Zeit im Feld.

sity and Jonathan joined the team, the writing up process was collective from the beginning. Given that, despite the nature of anthropological research depending on the anthropologist developing social relationships, collecting, and writing up ethnographic data is typically a solitary endeavor, this process itself created its own challenges and opportunities. Jonathan's first task within the group was to read through and then code Catherine's fieldwork notes. Carolin contributed to this systematic review of the fieldwork data by coding the interviews that Catherine had collected. Eveline monitored and coordinated coding the notes and interviews and made suggestions for their theoretical framing. While this work helped the team to better understand Barrio Logan through Catherine's lens, reading, looking at and listening to the data collected formed an incomplete picture, particularly for Jonathan and Carolin, who had not previously spent time in the South-Western United States. The field diary is vital in recording detailed information about the ethnographer's conversations, the things they have seen and events they have participated in on a daily basis. Reading through the field diary, looking through photographs taken and listening back to interviews when back from "the field" can also jog their memory, enabling them to read between the lines of the written page and remember extra detail that they may not have recorded at the time: for example, the smells, the heat, the emotions of the occasion, or further lines of a conversation. However, this process becomes obviously complicated by sharing this data with others. Ethnographic material is not objectifiable "data" that exist per se and could be collected, as it were – like fruit from the trees.[54] On the contrary, they are based on the specific social relationships that develop between the researcher and the research participants in each case and cannot be generalized. While this was known to the team, Eveline and Jonathan experienced the limitations of writing ethnography using another ethnographer's notes when working up drafts of articles for publication. Although Eveline, Jonathan and Carolin all combined their analysis of the data collected by Catherine with familiarizing themselves with other research on the U.S.– Mexico borderlands and Chicanismo, the joint fieldwork allowed them to relate to the data collected from their own perspectives of San Diego, and particularly Barrio Logan, including having conversations with some of the interlocutors, and in Carolin's case having directly conducted primary research herself.

Following the joint fieldwork, the manuscript itself was written in several loops. First, we created drafts of individual chapters, which we then edited and commented on – individually as well as collectively in meetings. Working in this way allowed us to exchange ideas for literature, to talk through different under-

---

54 Dürr, Feldforschung.

standings of our observations and to interpret them intersubjectively, to form arguments collectively and discuss their strengths and weaknesses, and all in all to support one another through the writing process. As we went along in this way, we also exchanged ideas with artivist Nanzi Muro who created the visualizations for the book. Gradually through this process the book began to emerge not as separate chapters but as an interconnected whole.

Chapter 3
"Sometimes you have to Transform into a Serpent": Political Subject-making around Chicano Park

At the center of social life in Barrio Logan is Chicano Park, where community meetings are held and where people come to from around San Diego, wider California and across the border to gaze at the murals on display, lowriders cruising by, and Aztec dancers practicing for ceremonial events. These *danzas* take place around the *kiosko* at the center of the park, which has been designed to resemble a Mesoamerican temple and was painted red, white and green, the colors of the Mexican flag. On a Sunday in early February 2020, Catherine was surprised to find approximately 100 Aztec dancers there, wearing full regalia: colorful tunics adorned with Mesoamerican symbols and feathered *tocados* (headdresses), rattle in hand, and ankle rattles. She joined the crowd to look around and understand what was happening. Soon, Mario, a tall young Mexican-looking man in a red t-shirt and a skinny beard approached her, asking, "Weren't you in the line at Las Cuatro Milpas yesterday?" Catherine recognized him, too, but was impressed that he would have remembered her from the popular, and therefore very busy, Barrio Logan restaurant. Mario told Catherine that the awareness that one needed to employ while dancing, including of one's surroundings, meant positively acknowledging those in one's vicinity, rather than just looking out for potential harm, and described this as a form of being *trucha*.[1] This reminded Catherine of a conversation she had had with the San Diegan professor of Chicana and Chicano Studies Roberto Hernández, who had explained that people are always *trucha* in the park, which is to say, there is always someone keeping an eye on who is coming and going there.

In a further conversation on being *trucha*, Mario recalled, "I think of an elder in a dance a year ago [...] Well, he fell and 'noticed' out loud that, 'Sometimes you have to transform into a serpent,' being on the vigil for positive outcomes instead of 'threats' was my thought." The elder's watchful behavior while dancing is an illustration of how users of Chicano Park produce it as a place through bodily interaction with it. Local knowledge, for example, about the correct way of moving

---

Note: A version of this chapter has previously been published as Alderman, Jonathan/Whittaker, Catherine: A Bridge That Divides: Hostile Infrastructures, Coloniality and Watchfulness in San Diego, California. In: *Sociologus* 71/2 (2021), pp. 153–174.

[1] For a detailed discussion of the term *trucha*, see Kammler, Trucha.

through a space, and local subjects are co-produced alongside one another as individuals move through a place. In this way, body and place interanimate one another.[2] Thus, Casey argues that "we are not only *in* places, but *of* them."[3] As Escobar has argued the things that we design in turn design us through their use.[4] In that sense, Chicano Park, through its users' interactions with it and within it, has a significant role in producing Chicanxs as subjects. In this chapter we show that being watchful – expressed by many Chicanxs as being *trucha* – has an encouraging, creative side to it that goes beyond being fearful of threats and surveillance.

Chicano Park was founded in 1970, a year after the Coronado Bridge was built over Barrio Logan, connecting mainland San Diego with Coronado Island (Fig. 5). The bridge evokes mixed feelings among the people of Barrio Logan, over which the bridge passes. On the one hand, the bridge facilitates access between different parts of the city of San Diego. However, it does so by passing through and over Barrio Logan, contributing significantly to air pollution suffered by local residents (see chapter 5). Following the demolition of a significant portion of what had been a thriving neighborhood for the construction of the bridge, the people of Barrio Logan asked the city council to build them a park below the bridge. Refusing to

Figure 5: The Coronado Bridge.

---

2 Appadurai, *Modernity at large*, p. 181; Casey, How to Get from Space to Place, p. 24.
3 Casey, How to Get from Space to Place, p. 19.
4 Escobar, *Designs for the Pluriverse*.

meet these requests, the city council were about to use the space for a highway patrol station without consulting local people, thus planning to add more police to an already overpoliced, criminalized neighborhood. Exclaiming "Ya basta – enough!" local people took matters into their own hands, taking over the space to collectively create a park themselves. These concerted political actions involved in the creation of the park became an important element in the emergence of residents of Barrio Logan and Logan Heights as political subjects – which is to say, as Chicanos and Chicanas. At the same time as demarcating a Chicanx space, we argue in this chapter that Chicano Park has been key to creating a sense of community through a particular aesthetic politics that incorporates the bridge.

In order to analyze these processes, we draw on anthropological approaches to infrastructure and the work of Jacques Rancière on aesthetic politics, police and subjectivation[5] as well as the decolonial analysis of Peruvian sociologist Aníbal Quijano.[6] As the meaning and use of infrastructures are often contested by their users, we contend that they are ideal sites for political action in the Rancièrian sense of action that disputes structural social arrangements that reinforce hierarchies, and makes claims for equality. We argue that through disputing the common sense of coloniality, Chicanxs emerged as what Rancière calls a "community of sense."[7]

We link this theoretical discussion to practices of watchfulness in Chicano Park and focus on individuals who are non-state actors, contrasting the surveillance of police or officials representing organs of the state.[8] We draw particularly on the testimonies of activists such as Damiano, a member of the Chicanx organization Unión del Barrio, who organizes the educational *Noches de Defensa y Resistencia* (Nights of Defense and Resistance) that take place near Chicano Park, and in which those in attendance remember the creation of Chicano Park and the history of the Chicano movement. As a historian in his late thirties, Damiano seeks to employ knowledge of the past to help shape the future of the park, while being critical of the Chicano Park Steering Committee's (CPSC) plans to create a museum next to the park. The CPSC, as we will go on to describe, arose among the local people who took over the space below the bridge in 1970, and manage the park in negotiation with the city government and other stakeholders. Alongside the Brown Berets of Aztlán, who guard the park, they are the park's stewards. The San Diego Brown Berets National Organization, are a revolutionary, semi-militant organization that,

---

5 Rancière, *Dissensus*; Rancière, Contemporary Art.
6 Quijano, Coloniality of Power.
7 Rancière, Contemporary Art.
8 Frekko/Leinaweaver/Marre, How (not) to talk about adoption.

like the African American Black Panther Party, emerged in the wake of the American civil rights movement of the 1960s;[9] some members of the CPSC are also active members of the Brown Berets. Another testimony stems from Carla, who was among the people who occupied the site of the park in 1970 and was involved in the Chicano movement from its very beginning and her earliest youth. Today, she volunteers as a tour guide of the park and informs visitors about the history and struggle of Chicanx people. Her description of events influenced this chapter considerably; it also includes Joaquín, a Chicano protector of the park who describes his and others' watchfulness there. As the son of Chicanx organizers, Joaquín was raised to serve the park, and does so in multiple roles: as a member of the CPSC alongside Carla, as a Brown Beret, and as a member of the Chicano Park Day's clean-up crew.

In the following theoretical discussion, we outline the conversations in anthropology around the topic of infrastructure to which this chapter contributes. Specifically, we discuss how this study advances perspectives on the coloniality of infrastructure to include the watchfulness engendered in individuals' relationships with it, through an analysis of the social relationship people in Barrio Logan have with the Coronado Bridge. Next, we discuss the Coronado Bridge's construction through Barrio Logan as an expression of coloniality and the people of Barrio Logan's ambivalent relationship with it. Then we look at how relations of coloniality were challenged through an aesthetic politics that incorporated the bridge into a Chicanx visual aesthetic. This leads us to discuss the very present danger that the bridge represents for those in the park below it. Finally, we return to the themes of being *trucha* and being serpent-like. Through the chapter, we show that residents of Barrio Logan are far from passive victims of discrimination, but active agents in the struggle to shape their community. We draw on the voices of local people to show how they challenge coloniality by subverting the original purpose of the bridge, and using it as a canvas for their memory work, re-designating the bridge as a Chicanx space.

Becoming Chicanx subjects under the Coronado Bridge

The physical grandeur of infrastructure projects such as the Coronado Bridge can lead them to be "depicted as barely changing behemoths."[10] However, the emergent

---

9 Palacios, Multicultural Vasconcelos.
10 Harvey/Jensen/Morita, Infrastructural complications, p. 11.

and experimental nature of infrastructure should be recognized.[11] On the one hand, large infrastructure projects can often justifiably be described as a "technology of liberal rule,"[12] which can be "a sociopolitical terrain for the reproduction of racism,"[13] elevating some, while marginalizing populations elsewhere.[14] However, following Brian Larkin's description of infrastructures not just as things but as the relation between things,[15] we can understand the characteristics of infrastructure as emerging out of its interactions – including with those populations ostensibly marginalized by it. That makes it unlikely that it will follow the plans of any single actor.[16] Taking note of the relational aspects of infrastructure underscores that infrastructure is above all the foundation for other activity.[17]

We bring the discussion of infrastructure in anthropology into conversation with Rancière's concepts of the police and politics to analyze the construction of Chicano Park by Chicanxs in Barrio Logan as a political act challenging the existing social order. Rancière introduces the "police" as a concept that does not describe an institution of explicit repression and control, "but a symbolic constitution of the social"[18] that Rancière refers to as the "distribution of the sensible," and that is enforced by a wide variety of institutional and non-institutional actors in society. The police order, according to Rancière, is the tacit understanding of what is and is not visible, doable, audible and sayable (etc.) within a given society to give the illusion of a commonly held consensus, and the inclusion of everyone of significance in public life. This consensus is reinforced through everyday aesthetic experience – the "distribution of the sensible." By contrast, "politics," for Rancière describes the opposite effect: actions that make "what was unseen visible" and "what was audible as mere noise heard as speech."[19] The Chicano Park takeover and painting of the murals by Chicanx artists that dispute the sensible and provide an alternative reading of their history and relationship to the border are examples of such politics. They create what Rancière refers to as "dissensus": the disruption of the taken for granted assumptions about the social order that are enforced particularly through aesthetics. The manifestation of dissensus is the reconfiguration of

---

11 Ibid; Alderman/Goodwin, Introduction.
12 Appel/Anand/Gupta, Promise of Infrastructure, p. 14.
13 Ibid., p. 2.
14 Howe et al., Paradoxical Infrastructures, p. 10.
15 Larkin, Poetics of Infrastructure, p. 329.
16 Harvey/Jensen/Morita, Infrastructural complications, p. 11.
17 Boyer, Revolutionary infrastructure, p. 175.
18 Rancière, *Dissensus*, p. 36.
19 Ibid.

"what can be done, seen and named" in a particular space.[20] Harvey and Knox argue that all infrastructure is political "not just though the transformations that they promise but also by arranging and rearranging the mundane spaces of everyday life"[21]; however, using Rancière's understanding of politics, not all infrastructure is political in itself, but interactions with infrastructure can be political when the political subject as such, those whose rights are not properly taken into account by those with political power, or as Rancière puts it "the part of those that have no part"[22] – acts to deliberately disrupt the established social order. Thus, infrastructures can be political through their "capacity to articulate the social in unexpected ways," fostering political communities around their "desire for, or rejection of, a particular infrastructural experiment."[23]

We can thus see that Rancière's concept of politics is particularly relevant in analyzing the struggles of the people of Barrio Logan in establishing Chicano Park because of the closeness he identifies between politics and aesthetics. Since the establishment of Chicano Park in 1970, artists have used the support structures of the bridge as canvasses for murals depicting a Chicanx subjectivity drawing on religious and historical figures significant in the collective Chicanx imagination. Rancière proposes that it is through provoking a dispute in the distribution of the sensible that community is formed.[24] Thus, as we will show in this chapter, drawing on history as seen through a Chicanx lens enables the people of Barrio Logan to create a sense of community that aesthetically challenges the coloniality of local experience in San Diego.

Further, we relate the building of the Coronado Bridge as infrastructure to the watchfulness and subject formation of Chicanxs. Following Larkin, infrastructures are interesting because they "reveal forms of political rationality that underlie technological projects,"[25] which enables practices of governmentality. Yet infrastructures are not just determined by their function; just as subjects are not exclusively governed by them. Infrastructure can be analyzed following their semiotics and emerging aesthetics in political reconfigurations. The location for the construction of the Coronado Bridge, and the lack of consultation with the people living in that location, demonstrates whose interests were being taken into account by the city's decision-makers, who do not themselves live in Barrio Logan, the area direct-

---

20 Ibid., p. 37.
21 Harvey/Knox, *Roads*, pp. 7f.
22 Rancière, *Dissensus*, p. 70.
23 Reeves, Infrastructures of Hope, pp. 715f.
24 Rancière, *Dissensus*, p. 37.
25 Larkin, Poetics of Infrastructure, p. 328.

ly affected by the bridge, but are able to benefit from the greater connectedness that the bridge creates between different parts of San Diego.

In this regard, attention towards infrastructures offers a rich area for the anthropological exploration of vigilance due to the possibility of defamiliarizing and rethinking the political dimension of technologies.[26] Although infrastructures are built upon a promise of modernity and equality for the people, the way they are performed continuously produces differences, and therefore contestations. As Hannah Appel, Nikhil Anand and Akhil Gupta suggest, "infrastructure is a terrain of power and contestation."[27] This means that infrastructure projects make visible the distribution of resources, which communities benefit from them, and which do not, forcing them to fight for necessary infrastructure provision.[28] Infrastructures are not just critical sites for analysis due to their role in the interaction between states and subjects. The way these relations are performed and contested opens a potential space for ethnographic research. One of the effects of such resource allocation will be discussed later in the book, in chapter 5, when we examine the pollution experienced in Barrio Logan as a combined effect of infrastructure construction around the neighborhood and its categorization as a mixed-use zone.

Citizenship-in-action: Building Chicano Park and creating community

The I-5 freeway was completed in 1963 and the Coronado Bridge in 1969, following the National Freeway Act in 1956. This infrastructure physically tore apart Logan Heights, reducing access on foot within the neighborhood; it was then that Barrio Logan as a separate neighborhood came into being.[29] The construction of the freeway and the bridge were together part of the "urban renewal" policy, which brought new highways, parking lots, public facilities and housing projects throughout the United States. While this gave rise to many impressive construction projects, it also threatened and evicted many communities, often communities of people of color.[30] This was also the case in San Diego. At the same time as connecting the mainland with Coronado Island, the building of the bridge meant the displacement of working-class families and businesses that had previously occupied the space. This is why the perception of this bridge differs in the overall context of

---

26 von Schnitzler, Temporalities of "Transition"; Appel/Anand/Gupta, Promise of Infrastructure.
27 Ibid., p. 2.
28 Ibid.
29 Galaviz, *Expressions of Membership and Belonging*, p. 20, p. 32.
30 Kirszbaum, Urban Renewal in the USA.

San Diego. While for the residents of Coronado Island it ensures unhindered access to the mainland and is considered a landmark of the city, for Chicanxs it is also a symbol of the unequal power relations within San Diego.[31]

In this context, Chicano Park was created because, in the words of Carla, it would allow residents of Barrio Logan to "maintain some sense of community." We argue that it was the creation of Chicano Park by the people of Barrio Logan *themselves* that reinforced a sense of community that fed into the collective identity of the community as *Chicanxs* among people with disparate origins. Gloria Anzaldúa has described Chicanxs as defined by the borderland in which they live, whose subjectivity as mestizxs includes elements of being Mexican, Native American and U.S. American.[32] An in-betweenness similarly characterizes the process of becoming a subject for Rancière, for whom political subjectivation relies not so much on any primordial common identity, but on the "the enactment of equality – or the handling of a wrong – by people who are together to the extent that they are between."[33] Group formation, as Barth has acknowledged, can have a primarily oppositional character to it.[34] In Barrio Logan, although Chicanx organizations emerged around a narrative of common belonging to the locality, they included people whose families had lived in San Diego for generations, as well as recent migrants, as viewed from the perspective of the state. Chicanxs often challenge this division of their community between "U.S.-born" and "recently arrived" people, instead foregrounding the artificiality and coloniality of the border that does not simply mark but has in fact created that division.[35] Transcending such divisions, Chicanxs struggle against the discrimination that racialized people face on a daily basis and which is reinforced through the structures that privilege white middle-class San Diegans over working-class Mexican Americans and other racialized and migrantized people in the U.S.

Due to the people of Barrio Logan's persistence, city authorities had promised them a 1.8 acre space, albeit directly below the overpasses to the Coronado Bridge.[36] However, as Carla explained, on 21st April 1970, a student and resident walking through the area noticed "bulldozers that were leveling the land. When they were leveling he got very excited, because he thought they were finally gonna build the park." However, when he realized that he was mistaken, "he runs all the way through the community to some of the older residents, letting

---

31 Kühne/Schönwald, *Eigenlogiken*, p. 207.
32 Anzaldúa, *Borderlands/La Frontera*.
33 Rancière, Politics, Identification, and Subjectivization, p. 61.
34 Barth, Introduction, p. 11.
35 Hernández, *Coloniality of the US/Mexico border*.
36 Galaviz, *Expressions of Membership and Belonging*, p. 33.

them know that this was not gonna be a park, it was gonna be the California Highway Patrol substation."

Manuel Galaviz writes in his work on Chicano Park that the initial protest was poorly attended,[37] but that "The following day (April 22nd, 1970) when community members alongside local middle school and high school students got involved in the political action they were able to form human chains around the construction site."[38] As Carla explained:

> Anyway, we surrounded the bulldozers, so, [...] we stopped them [and] we began planting our own park. That night [...] we formulated the CPSC, who are the stewards of the park. We formulated in order to negotiate with the city and to be there because now we have stopped the State of California and the city of San Diego from doing their work. We all occupied the land for twelve days, and then negotiations began to actually paint this park. And for the city to swap the land for that California Highway Patrol substation. And, we began working on the park at that time.

In taking the land for their own purposes, the community were "acting out of the presupposition of equality."[39] This is, for Rancière, the very definition of a political act: an action which is performed by the *demos* themselves, out of the assumption of their own equality as a group with those wielding greater power, rather than waiting for equality to be distributed to them.[40] It is what Rancière regards as the most democratic form of expression. For Rancière, politics concerns "the part who have no part," disrupting the police order that excludes or marginalizes them, through the assertion of their own equality in that order, and in so doing demonstrating its contingency. In Beatriz Zamora's children's book, "The Spirit of Chicano Park," the CPSC co-founder and chairwoman-for-life, Tomasa Camarillo, is quoted as saying: "Chicano Park gives us hope. The day we stood up for the park, we found our voice! We told the world that we mattered and would have our park, even if we had to build it ourselves!"[41]

After twelve days of occupation, the City Manager agreed to begin negotiating if the protesters left the site. On 31st June 1970, a $21,841.96 contract to develop the space into a community park was commissioned.[42] Although maintenance of the park was officially overseen by the city's Parks and Recreation Board, since

---

37 Ibid., p. 34.
38 Delgado, A Turning Point; Cockcroft, The Story of Chicano Park, cited in Galaviz, *Expressions of Membership and Belonging*, p. 34.
39 May, *Political Movements*, p. 20.
40 Ibid., p. 21.
41 Zamora, *Spirit of Chicano Park*, p. 31.
42 Delgado, A Turning Point.

then it has been the CPSC who have practically looked after the park, which they refer to as a "the story of a barrio tragedy transformed into triumph."[43] It can be seen to represent what Lemanski refers to, in reference to the adaptation of infrastructure by citizens to better suit their needs, as "Citizenship-in-action."[44]

Bridges are commonly conceived as connecting otherwise disconnected places,[45] and bridges as "passageways, conduits and connectors that connote transitioning, crossing borders, and changing perspectives" have also been used by Anzaldúa[46] as a figurative symbol for the Chicanx condition. Bridges, for Anzaldúa, span the liminal, in-between space that she calls nepantla. In This Bridge Called My Back (2002), edited by Cherrie Moraga and Gloria Anzaldúa, the bridge symbolizes the collaborations of Women of Color to share their common experiences of struggle at the intersections of race, gender, sexuality and culture. Complicating this metaphor, the Coronado Bridge has created division, as much as connection. Where once there had been a large, vibrant neighborhood, noisy overhead freeways supported by mounds of grey concrete physically split the community and caused considerable aesthetic degradation. At the same time as disconnecting Barrio Logan from Logan Heights, the neighborhoods most affected by the pollution directly caused by it, the Coronado Bridge passes over Barrio Logan and connects the mainland with Coronado Island, a highly-privileged, wealthy, majority white community by the beach and the location of San Diego's U.S. Navy presence. In another sense, however, the bridge is a connection point between people with disparate identities, not just because of its aesthetic presence, but because a political community has formed against the coloniality that the bridge represents.

Since its construction, the people of Barrio Logan have reflected on the ambivalent character of the bridge and its impact on the ways they see themselves. On one of the educational tours that the CPSC offers in Chicano Park, the guide, Carla, described it as a "beautiful bridge," but also emphasized that it had devastated the community. According to Carla, the bridge has been the direct cause of the population dropping from 20,000 to 5,000 in less than ten years. In a similar vein, a mural in the Barrio Dogg diner depicts the Coronado Bridge under a banner of Barrio Logan, sublimely enhanced by a stunning setting sun (Fig. 6). When asked about this, the owner clarified that it is not his intention to glorify the bridge, but that the bridge is on his wall because of its undeniable presence in Barrio Logan. In this case, infrastructure is not that which is invisible,[47] but rather claims

---

43 Anguiano, Battle of Chicano Park.
44 Lemanski, Infrastructural citizenship, p. 596.
45 See, e.g., Simmel, Bridge and Door, p. 10.
46 Anzaldúa, *Borderlands/La Frontera*, p. 1.
47 Star, The Ethnography of Infrastructure, p. 380.

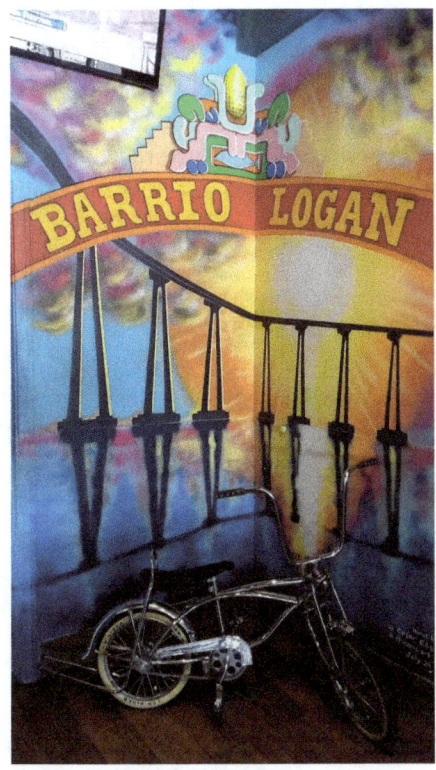

Figure 6: The mural of the Coronado Bridge in the Barrio Dogg diner.

an overbearing and inescapable presence. Thus, while the bridge can be viewed as an aesthetic reminder of inequality, a symbol of the neglect, disrespect and oppression that Chicanxs in San Diego have endured, and a very visible reminder of the state's presence and its priorities,[48] besides itself creating a dangerous environment in which personal watchfulness is a necessity, it has nevertheless come to be a physical symbol of Barrio Logan that local people themselves identify with.

Infrastructure projects, Harvey, Jensen and Morita argue, are not "unmitigated public goods," but "turn out to be good for some people, some of the time."[49] This is because they reproduce the visions of their funders, designers and the politicians in power at the time. As becomes apparent in our case study, infrastructure can be thought of as a "technology of liberal rule,"[50] which can be a sociopolitical terrain

---

48 Lemanski, Infrastructural citizenship, p. 592.
49 Harvey/Jensen/Morita, Infrastructural complications, p. 8.
50 Appel/Anand/Gupta, Promise of Infrastructure, p. 14.

for the reproduction of racism.⁵¹ As such an infrastructure project, the Coronado Bridge was a physical manifestation of coloniality, reinforcing structural racism that people in Barrio Logan are conscious of on a daily basis, and is acknowledged in the way that racialized people move about the city. People in Barrio Logan also responded aesthetically to the coloniality of the Coronado Bridge, through murals that highlight Chicanx and Raza decolonial struggles locally and across the continent.

The aesthetic politics of Chicano Park's murals

The painting of the pillars of the bridge began in 1973 when Victor Ochoa, an internationally renowned Chicano artist, orchestrated a "mural marathon," inviting Chicanx artists from all over the Southwest to come and paint.⁵² Since then residents and student groups of the neighborhood have added to the murals in Chicano Park after their designs were first approved by the CPSC. Following in the footsteps of the Mexican muralists, the murals pursue educational goals and depict the genealogy and heroic figures of the Chicano movement, picking up motifs from different historical periods, cultures and events and directing their messages both to Chicanxs themselves and to outsiders.⁵³ The murals serve to raise political awareness and provide a continuous reminder to be vigilant against encroachment, as they show land grabbing and overreaching as inherent features of Chicanx history. The aim is to convey the history that is excluded in state schools in order to achieve a sharpening of political awareness, which ultimately serves as the key to success for the continuation of the "struggle." Through these messages, Chicano Park has become a place of education and identity that encourages the continuation of collective political struggle in the present and to map the goals and achievement of the movement. We argue that in itself, the struggle to create Chicano Park has been a significant aspect of the development of Chicanxs as "subjects of struggle."⁵⁴

The murals as a whole make a claim to Chicanx belonging that is at the same time a challenge to dominant and exclusionary understandings of "America" as

---

51 Ibid., p. 2.
52 Josie Talamantez of the Chicano Park Steering Committee identified 49 murals created between 1970 and 1989 when she did the national register listing of the park's murals (Talamantez, *Chicano Park and the Chicano Park Murals*, pp. 43–67, Latorre, *Walls of Empowerment*). There is further information available at e. g., https://digital.sandiego.edu/ethn-books/2 and https://chicano-park.com/cpmap.html.
53 Cf. McCaughan, The Border Crossed Us.
54 Gutiérrez Aguilar, *Horizonte Comunitario*.

Figure 7: Mural commemorating the creation of Chicano Park and memorial to the people killed in Chicano Park.

limited to the United States of America (rather than the continent) and that privilege white U.S.-American perspectives of what it means to be an American. The murals combine depictions of a pre-colonial history with illustrations of more recent local and wider Raza decolonial and anti-imperialist struggles; some blur temporalities, making direct connections between the current political movement and pre-colonial forebears. One mural Carla pointed out to the tour group was of the student who had run around informing the community about the bulldozers (see Fig. 7). He is depicted as a Mayan runner, carrying a torch. Other murals reference a wider de-colonial and anti-imperialist project in the Americas. There are murals in which Fidel Castro and Ernesto "Che" Guevara appear; Carla pointed out another of Emiliano Zapata, who "fought with Pancho Villa in the Mexican-American war"; as Carla put it, "he'd rather stand on his feet than – or die on his feet, than live on his knees." The heroic portrayal of figures who have been antagonistic to U.S.-imperialism in Latin America, such as Fidel Castro, or who have actively fought against the U.S., such as Zapata and Pancho Villa, challenges Anglo-American narratives of Raza figures as a menacing others, rather than as heroes, and of the need for a strong border to keep back potential threats to the nation.[55] The border itself is further questioned through murals that emphasize Chicanx forebears' mobility through time, such as one that portrayed the founding of Mexico City, as Carla explained,

> it was the Aztecs that were coming from the North, going to the South. Interesting, huh, cause everyone's going from South to North now. [Laughs] But they were looking and their spiritual visions had told them to look for a lake that had in the center a little piece of land with a cactus and a, a cactus with an eagle with a serpent in its mouth. And that's kind of the center of the Mexican flag. And that was the founding of Mexico City. That's what this mural is.

The migration from what is today the southwestern United States into Mexico by Chicanx' pre-colonial forebears is connected to what Chicanxs understand as Aztlán, which is discussed in detail in the following chapter.

There are also murals of the Earth goddess Coatlicue and the Virgin of Guadalupe. As indicated in the introduction to this chapter, the serpent is a significant figure for Chicanxs, because it connects them to their pre-colonial ancestry. Anzaldúa writes of Coatlicue, also known as Tonantsi, or Tonantzin, a Mesoamerican fertility goddess, as having "a human skull or serpent for a head, a necklace of human hearts, a skirt of twisted serpents and taloned feet."[56] Coatlicue's face is composed of two fanged serpents and her skirt is of interwoven snakes. In one mural, Coatlicue (Fig. 8) is painted on the back of a mural depicting the Virgin of Guadalupe, which gestures to the historical continuity of worship between the two. The Virgin of Guadalupe first appeared in 1531 on the spot where the Aztec goddess had been worshipped and told Juan Diego, who first saw her, that her name was Maria Coatlalopeuh, which Anzaldúa interprets as meaning "the one who is at one with beasts," *coatl* being the Nahuatl word for serpent and *coatlopeuh* meaning "the one who has dominion over serpents."[57] Anzaldúa further writes that the

> Virgen de Guadalupe is the single most potent religious, political and cultural image of the Chicano/mexicano. She, like my race, is a synthesis of the old world and the new, of the religion and culture of the two races in our psyche, the conquerors and the conquered. She is the symbol of the mestizo true to his or her Indian values […]. To Mexicans on both sides of the border, Guadalupe is the symbol of our rebellion against the rich, upper and middleclass; against their subjugation of the poor and the indio […]. She mediates between the Spanish and the Indian cultures […] and between Chicanos and the white world. She mediates between humans and the divine, between this reality and the reality of spirit entities.[58]

---

55 Chavez, *The Latino Threat*.
56 Anzaldúa, *Borderlands/La Frontera*, p. 27.
57 Anzaldúa, *Borderlands/La Frontera*, pp. 28f.
58 Ibid., pp. 30f.

Guadalupe/Coatlalopeuh is thus iconic for Chicanxs, and a popular clothes and tattoo design among young community members, because she is herself a *mestiza* who connects them with their Indigenous ancestry.[59] According to Anzaldúa the church of the Spanish "desexed *Guadalupe*, taking *Coatlalopeuh*, the serpent/sexuality, out of her,"[60] creating a division between Guadalupe as a chaste virgin and Coatlicue amongst other Indigenous deities as the work of the devil. Her analysis gives us a sense of why the serpent has been reclaimed as a key symbol in contemporary Chicanx decolonization struggles. However, as discussed in chapter 1, Anzaldúa has also been criticized for this characterization, as it erases differences in experience, particularly with regard to discrimination and coloniality, between *mestizxs* and Indigenous people.

The murals present a threat to the dominant social order by disrupting "the sensible" of commonly held assumptions of acceptable aesthetic expression within

Figure 8: The mural of Coatlicue.

---

59 Ibid., p. 27, p. 30.
60 Ibid., p. 27.

U.S. society, powerfully displaying an alternative reading of historical events and current social relations. This has become evident when, over the years, the murals have been attacked by white supremacists, and in 2018, white supremacist marchers threatened to paint over the murals. Under particular attack have been murals of the Earth goddess Coatlicue and representations of Fidel Castro and Che Guevara. Indicating the mural of Coatlicue, Carla pointed out the drippings of red, blue and purple paint, which had been thrown from the bridge by KKK members before a Chicano Park celebration:

> they threw paint on the kiosk that landed on the pill- on the stabs, well, that was easy to paint. But they threw paint on this mural here. And this artist, Susan Yamagata, decided to incorporate it into her mural. And leave the drippings, so that no one would ever forget that these murals were attacked and that this park and this community was attacked.

Diego, an American Indian movement activist who frequently collaborated with Chicanx organizations, described to Catherine on a separate occasion how in recent years the interference of Proud Boys (an armed right-wing militia movement) and other white supremacists in Chicano Park has included demanding that the flag of Aztlán (see chapter 4) be taken down, and the name of the park be changed, because according to them it is an "American park." They also demanded that some murals be painted over and the statue of Pancho Villa be removed as he was a "Brown supremacist." In response to this latter incident, according to Diego, in 2016, around 800 local people turned up to defend the park. Police also arrived on the scene, supposedly because in response to the threats made by the white supremacists, but according to Diego, Chicanxs were wary of the police's appearance on the scene because of their own experiences of racist discrimination at the hands of the police.

The murals in Chicano Park demonstrate the power of urban popular art, which as Oosterbaan and Jaffe have argued, has the ability both to enforce consensus and to force a break from it.[61] The murals in Chicano Park are threatening to white supremacists because they allow marginalized people to express their own perspective on the world. In so doing, they disrupted white supremacists' aesthetic perspectives on what they perceive as "America," creating what Rancière calls dissensus.[62] By vandalizing these images, white supremacists affirm "the police" as the order of the visible and the speakable, while the murals act as a challenge by disrupting the accepted distribution of the sensible. Rancière has pointed out that the police order is not simply enforced by the institution of the police itself,

---

61 Oosterbaan/Jaffe, Popular Art.
62 Rancière, *Dissensus*, p. 69.

but also depends on how the sensible is distributed and shared: "a matter of how the sensible is distributed, partitioned and shared."[63] This is why periodically white supremacists come to Chicano Park and threaten to destroy the murals: because murals that proudly display images of figures that many right-wing and nationalistic U.S. Americans have antipathy to make visible alternatives to the social order that is underpinned by coloniality.

Because of these looming threats to the park, its guardians and other Chicanx community members have to be constantly alert over what kind of activities are taking place in the park. This involves also an ongoing negotiation between the CPSC and both users and the Parks and Recreation Board, which is an advisory board operating at the level of the whole of the city of San Diego. Such negotiations show that even after political struggle has created new common space it continues to entail ongoing negotiation between actors with distinct perspectives and interests.[64] Anyone who wishes to paint a new mural must first consult the CPSC who will then work with the artist to select a pillar and ensure that the design reflects the theme of the park. The CPSC are particularly cautious in allowing any activities to take place that may make money from the location and its history on display. In 2020, as Catherine learned at a CPSC meeting (see chapter 2), they turned down a request to shoot a documentary in the park because it was not clear what it was about, and when it was reported to them that a car commercial was being shot in the Park, they succeeded in stopping the filming. The CPSC's attention extends to more mundane uses of the park. As Carla was leading her tour group, she admonished someone running through the park: "You're not supposed to be running through the park! The grass is all seeds!" As researchers, we ourselves experienced policing of unacceptable behavior within the park, when Jonathan took a photo of a graffiti related to gang activity and was immediately approached and told to delete the photo by a presumed member of the local *clika* (gang). This made us aware of the fact that we were watched closely while on fieldwork in the park, while we were observing activities ourselves and documenting the built environment.

However, negotiation over the use of the park can be particularly contentious between local residents themselves. In 2010, a proposal was presented to the CPSC for the construction of a Veterans' Memorial to honor U.S. military service people from Barrio Logan. The discussion became one involving controversial symbols of U.S. patriotism and imperialism in a predominantly Chicanx and Mexican American neighborhood.[65] After a year of debate between the Logan Heights Veterans

---

63 Ibid.
64 Postero/Elinoff, A return to politics, p. 6.
65 Galaviz, *Expressions of Membership and Belonging*, p. 38.

Memorial Committee and CPSC members over the issue, the final design took into account the requests of the CPSC by resembling a Mesoamerican geometric pattern.[66] At the heart of the controversy over the memorial's design was the inclusion of a flagpole flying the U.S. flag, which the CPSC felt clashed with the aesthetics of the park that were otherwise rooted in an opposition to U.S. imperialism and colonialism. The process called into question who had ultimate power over what occurs in the park and the messages on display. Through the murals, what emerged among Chicanx protectors of the park (the CPSC and the Brown Berets) was a "a frame of visibility and intelligibility that puts things or practices together under the same meaning, which shapes thereby a certain sense of community," what Rancière calls a "community of sense."[67]

However, their negotiations over the aesthetics of the park with other local residents show that disagreements exist between neighbors over the character of the park, with veterans' groups defending a partition of the sensible that promotes consensus around symbols of American patriotism. Further, Chicano Park is promoted as a tourist destination and has been designated as a national historic landmark since 2017.[68] In this vein, Chicano Park has become part of San Diego's tourism economy and will be integrated in its marketing strategies. For instance, the San Diego Trolley Tours have added a new stop in Barrio Logan in 2022. It can be expected that these developments will further accelerate the ongoing gentrification in the Barrio, which adds to threats of displacement and involuntary relocations. As we will explain in the next section though, users of the park are also required to engage in watchfulness from within the park of threats from the bridge, which in a very real sense has the potential to be hostile infrastructure, posing a danger to all those below.

Imminent danger from the bridge

The bridge, however, is not just a symbol of state neglect, but an actual ever-present danger. The CPSC asked the City Council to secure the edges of the bridge after cars had previously fallen off, yet the city did nothing. Then in 2016, a speeding car, driven by a Navy Petty officer, drove off the bridge, killing four people in Chicano Park below. According to a report in The San Diego Union Tribune, the

---

66 Ibid., pp. 45–48.
67 Rancière, *Dissensus*, p. 31.
68 Kühne/Schönwald/Jenal, Bottom-up memorial landscapes, p. 5; San Diego Tourism Authority, n.d.: https://www.sandiego.org/articles/parks-gardens/chicano-park.aspx [last access: 07/11/2022].

driver was found guilty of vehicular manslaughter and driving under the influence. After drinking earlier in the day, he was accused of losing control of his pickup, while allegedly arguing with his girlfriend and speeding (81 mph) at the same time as trying to pass another car on his left.[69] The tour guide, Carla, recounted how that day was the day of a motorcycle run from Los Angeles that ended in Chicano Park. The park was therefore busy with people with motorcycles and vendors with selling merchandise. A memorial was later created to the four people who were crushed and killed when the car left the bridge, which is maintained periodically by their families. Two of the people killed were from Los Angeles and another two from Arizona, and many local people were injured.[70] Despite being sentenced to nine years and ten months in prison, the driver was released early in November 2020, having served two years and ten months.[71] In response, in a letter to California governor Gavin Newson dated 4th November 2020 (posted on Barrio Bridge Facebook page)[72], the CPSC expressed their outrage at the approval for early release and explicitly connected the decision of the California Department of Corrections and Rehabilitation (CDCR) to a longer and wider pattern of disrespect towards the communities around the site of the bridge. They cited the "[destruction of] our neighborhoods by placing freeways and bridges down the middle of our homes and businesses" as context for the deaths of the four people below the bridge.

The letter conveyed a sense among the people of Barrio Logan that this incident follows a "historical pattern" of neglect, violence, and silencing. It had not been the first time that the law had not been on their side and they had felt treated as second-class citizens. They argued that the decision to release the driver early "speak[s] directly to a larger historical pattern of unequal justice for Chicano and other communities of color," including the actions of the state of California "when they decided to come into our community over fifty years ago and destroyed our neighborhoods by placing freeways and bridges right down the middle of our homes and businesses." Thus, Chicanxs did not expect the law to work the way it is purportedly supposed to, as a fair arbiter between citizens and the state, but they had no choice except to appeal to it anyway. This case shows that through its very malfunctions, infrastructure can demonstrate who and what matters in the prevailing social order[73] as well as highlighting the bridge as a "state effect,"[74]

---

69 Repard, Driver Who Crashed Off Coronado Bridge, Killing Four, Found Guilty of Manslaughter.
70 Kucher/Figueroa, Driver who killed 4.
71 Ibid.
72 https://www.facebook.com/298152410237632/photos/a.298647426854797/3699360696783436/ (11/04/2020).
73 Chu, When infrastructures attack, p. 353.

that is as a material manifestation of the state. The latent danger of cars potentially falling from the bridge and physically violating Chicano Park as a safe space, echoes the arbitrariness of the state's insertion of the bridge into Logan Heights. While the park has become a space for Chicanxs to express themselves qua Chicanxs, the presence of the bridge is a constant reminder that since their incorporation within colonial states (first Mexico, then the U.S.), to be Chicanx means to be under a continuously looming threat from outsiders that requires permanent alertness. Therefore, Chicanxs need to be *trucha*.

"Ponte trucha"

When Catherine met Mario at the Aztec *danza* ceremony in Chicano Park that we describe at the beginning of this chapter, she remarked to him that it was interesting to see such dances in a space where four people had died. Mario responded that his *calpulli* (*danza* group) was a defiant affirmation of life in a space of death, a warrior dance, as some poses and regalia demonstrate. Both the collective and individual actions of dancers and the park's guardians express a watchfulness over the park as a spatial expression of the subjectivity of the people of Barrio Logan as Chicanxs. This subjectivity includes an awareness of constant imminent danger, which Mario described through the alertness of the elder who declared that one had to "transform into a serpent."

This embodiment of the serpent may then be reflected back towards the bridge itself. In his study on Chicano Park, Galaviz includes a photo of an artwork by Hector Villegas in which the bridge morphs into the feathered serpent Quetzalcoatl.[75] The artwork highlights the ambivalent nature of the bridge, which in being incorporated into Chicanx organizing, becomes itself hybrid. Like the justice system which the CPSC describe as working against Chicanx interests, the bridge is unavoidable. As such, the people of Barrio Logan have attempted to work though the bridge. As a hybrid, the parts of the bridge that cross Barrio Logan become like the Virgin de Guadalupe. Neither one thing nor the other, but both.

The watchfulness around the park, understood as being *trucha*, is multifaceted. As outlined in the introduction, on the one hand it is a product of coloniality, on the other hand it contains the potential for decolonization.[76] What we show throughout this book is the significance of this attentiveness to one's surroundings

---

74 Mitchell, Society, Economy, and the State Effect; Harvey, The Materiality of State Effects.
75 Galaviz, *Expressions of Membership and Belonging*, p. 70.
76 Whittaker/Dürr, Vigilance, Knowledge, and De/Colonization.

in Chicanx individual and collective actions. As our interlocutors told us, they consider watchfulness as shaping their everyday life and describe it as an essential part of their sense of self. Watchfulness is largely fed by distrust of the Anglo-American way of life and worldview, which in our conversations was referred to as the "gringo way," and read as the Chicanx response to racist structural features of U.S. society. The injunction to always be on guard is passed down from one generation to the next – often as a major theme within the family.[77]

Even though it is clearly reflected by Chicanxs that vigilance cannot be separated from experiences of discrimination and, starting from basic fearful assumptions, can be emotionally stressful, there is also a positive connotation of vigilance in the Chicanx self-image. It is not primarily a matter of anticipating possible danger, but of an everyday, intense, above-average power of observation and perception of the people around oneself. This can manifest itself through attentive, approachable, and respectful interaction with other people, especially from one's own community. In this context, *ponerse trucha* means becoming attentive with positive connotations. In the context of Chicano Park, our interlocutors affirmed that the *trucha* "ethos" "is probably lived through our inability to get the city of San Diego to do what we want them to do. So many of our negotiations require for one to be *trucha*."

Anzaldúa describes the attentiveness of Chicanxs in similar terms using the term "la facultad." *La facultad* she portrays as a quick perception that sees deeper than surface reality. Someone "possessing this sensitivity is excruciatingly alive to the world,"[78] able to quickly sense anything that will break their defenses. This is the alertness described at the beginning of this chapter as possessed by the dancer, alert to his surroundings, and for whom through such alertness one transforms into a serpent. That is, like the serpent, a part of the Chicanx-mestizo condition, which Anzaldúa argues connects them with their Indigenous ancestors. A watchful awareness of their surroundings is a significant aspect of Chicanxs' subjectivity as Chicanxs, as watchful subjects, who need to be alert to their surroundings and particularly the dangers that may confront them.

Chicano Park requires permanent watchfulness precisely because as an aesthetic anticolonial project it is disruptive by nature. By mixing various elements of Raza culture, including Mexican, Latin American and Indigenous American symbols, Chicano Park represents an idea of being Chicanx that is not monolithic, does not always speak with one voice, but which makes space for personal iden-

---

77 See Dürr/Whittaker, Wachsamkeit als Alltagspraxis.
78 Anzaldúa, *Borderlands/La Frontera*, p. 38.

tifications.[79] Aesthetically, it challenges ideas of belonging and temporality within the United States (see chapter 4).

It is the park's disruptive nature that creates the need for watchfulness by its protectors. This practice, in turn, contributes to solidarity between the park's users and guardians, as they take care of the park collectively through "informal chains of alertness."[80] In doing so, Chicanxs in Barrio Logan challenge both the spatial and temporal order of dominant U.S. narratives of history, victory and "civilization" by projecting their own subjectivity through images of Raza heroes and an empire that pre-dates the borders set by both Spanish and American colonization. Such representations critique and undermine what Aníbal Quijano calls the coloniality of power, which was constituted materially and subjectively through a new space/time regime in the Americas referred to as "modernity."[81] The murals combine with individual and collective practices of watchfulness through which Chicanxs have emerged as a community around the park. In the next chapter, we inquire further into the particular kind of watchfulness we observed in Chicano Park as a way to protect the Chicanx spiritual and political project of Aztlán.

---

79 Cohen, Personal Nationalism.
80 Goldstein, *Outlawed,* p. 123.
81 Quijano, Coloniality of Power, p. 195.

Chapter 4
Watching out across Time and Space in Aztlán: Chronopolitics in Chicano Park

On a weekend in early March in 2020, many visitors gathered in Chicano Park to participate in a citywide architecture weekend. Joaquín, a young man in his mid-20s, with soft but serious features and short-cropped black hair and a mustache, dressed in a button-up, oversized lumberjack shirt typical of *Cholos*,[1] took advantage of this gathering to recruit some volunteers, including Catherine, for the upcoming Chicano Park Day celebrations. While he was watching out for potential helpers, he observed police officers approaching a group of unhoused people – presumably to order them out of the park and possibly harass them in the process. Joaquín recognized the potential danger that they might be harassed and, based on his experiences, suspected an arbitrary, inappropriate act of police violence. This is not uncommon in Chicano Park, and Joaquín had also previously observed people being forcibly removed from the park. As a budding community leader and a member of the San Diego Brown Berets National Organization, known for their militancy and ambivalent relationship with law enforcement,[2] he feels a special responsibility to maintain and protect the community, and decided to intervene. Joaquín slowly approached the scene. Following the Brown Beret ethos of "Observe, serve, protect," Joaquin did not want to escalate the situation, so he leisurely sauntered toward the group with his hands in the pockets of his baggy jeans and asked what is going on. The police officers reacted to his intervention in an aggravated manner precisely because Joaquín hid his hands in his pockets, and they suspected a possible weapon, one of the bystanders suggested. After being rebuffed by the police, Joaquín felt shaken and outraged by their presumptions to consider him – groundless, in his view – armed and dangerous.

In this chapter, we use this incident in Chicano Park as a prism through which to elaborate on the nexus of watchfulness, subjectivation and temporality and the spatial embeddedness of these processes. In doing so, we show how Aztlán is enact-

---

Note: A version of this chapter has previously been published as Dürr, Eveline/Whittaker, Catherine: Wachsamkeit als Alltagspraxis. Dekolonisierung von Zeit und Raum im Chicano Park in San Diego, Kalifornien. In: Brendecke, Arndt/Reichlin, Susanne: *Zeiten der Wachsamkeit.* Berlin 2022, pp. 179–210.

1 A *Cholo*, in the U.S.-Mexico borderlands, is someone with a particularly stereotypical Chicano way of dressing and attitude. As with "Chicano" itself, "Cholo" can be considered a slur in some contexts but has been reclaimed by many individuals as a self-identifying term.
2 Cf. Palacios, Multicultural Vasconcelos.

Open Access. © 2023 the author(s), published by De Gruyter. This work is licensed under the Creative Commons Attribution 4.0 International License. https://doi.org/10.1515/9783110985573-007

ed in Chicano Park. Aztlán is a socio-spatial concept referring to the region in the Southwest where the Aztecs are believed to have migrated from on their way to present-day Mexico. Thus, Aztlán constitutes ancestral land, conveying belonging for Chicanxs today.[3] It also has sacred-spiritual dimensions, which challenge the ongoing coloniality inscribed in hegemonic understandings of time and space. Drawing on practices and speeches of members of the Brown Berets, we highlight the ways they overcome Anglo American views of the past and present, by asserting their own notions of time and space in Aztlán. We further show that in these practices and speeches Chicanxs mobilize an affective relationship with Aztlán, which is tied into care and stewardship.

We argue that watching out over the anticolonial space framed as Aztlán requires guardianship and caring, which turns watchfulness into a specific social practice and elicits action. We contend that politicized subjects emerge through these practices, a process ultimately helping to decolonize and empower the community. However, we also discuss ruptures and frictions in this community, some of which run along the fault lines of older and younger Chicanxs. The older generation, for whom Aztlán served as a key concept to form a politicized collective in the 1970s, strives to keep the struggle going and to responsibilize the younger generation to speak out against coloniality and tackle current injustices. We argue that temporality is key in these chronopolitical messages, often framed with a specific urgency, which does not allow for hesitation or delays. In other words, this chapter shows how Chicanxs ensure that community members remain involved in the struggle and stay connected across generations.

Watching and being watched as political subject formation

In the scene described above, multiple regimes of colonial and decolonial vigilance were at play. Because of the special event and social gathering in the park, Joaquín himself was aware of the fact that he was being watched – as were the police. Thus, Joaquín most likely expected that his intervention would be closely followed, commented on, and evaluated by those present in the park. And indeed, while some of his Chicanx peers agreed with him, showing understanding for his displeasure, and found the behavior of the police disrespectful, others remarked that Joaquín still lacked experience. In their view, this was clearly the case because otherwise he would not have approached the police with his hands in his pants pockets, especially since, based on his phenotype and his clothing, Joaquín could

---

3 Anaya et al., *Aztlán: Essays on the Chicano Homeland.*

be read as a young, strong Mexican and thus racialized and criminalized by the police as a potential radical leftist or gang member. Because of his lack of foresight, in their view, Joaquín still counted as a learner for them. Thus, Chicanxs' cautious observation in the park is directed not only to the "outside" for potential threats, but also to the behavior of their own group members.

Joaquín's careful observation of the police, anticipation of their actions, and intervention can be read as an act of resistance in the sense of the Chicano movement, whose goals point beyond him as an individual and focus on combating injustices more broadly. Joaquín did not have a direct personal relationship with the unhoused but perceived them as persons who have to fear discrimination. Joaquín's act confirmed his prominent role in the collective as a Brown Beret and community leader. At the same time, he protected what he saw as disadvantaged and marginalized individuals from encroachment. In doing so, Joaquín publicly and performatively identified with the historical and continuing goals of Chicanismo and was able to recognize, politicize, and ultimately intervene at the right moment for his actions. With his commitment to *la Causa*, he visibly and *coram publico* connected to the struggle of his predecessor generation and continued to write the community history depicted on the park's murals (see chapter 3). In doing so, Joaquín's transformed a grim reality into an alternative Chicanx reality by intervening on the spot and without hesitation.

Further, we interpret this intervention as revealing the relationality of subjectivation, sociality and temporality. Joaquín, through his interference under the eyes of the audience in the park, takes up the expectation placed on him as an active Chicano and Brown Beret to stand up for the goals of the movement and puts this performatively into action. By approaching the police in this scene, he stepped out of the ascribed social position of the discriminated subject and demanded justice. This can be read as an act of empowerment of a racialized, quasi-colonial subject in the sense of Frantz Fanon[4], who discusses the potential of colonized subjects to free themselves from the effects of the "white gaze." From the perspective of Rancière[5] we can interpret Joaquín as acting in a way that presupposes the equality that is ordinarily denied to him. Through these practices, Joaquín becomes a political subject and exercises agency. Joaquín, as a potential suspect, claims the position of a guardian and caretaker through his action. Watchfulness is inherent to each of these social positionalities. It is now up to Joaquín to manage the police and state power to ensure that there are no encroachments by dominant forces in the park. Further, this act of subjectivation shows the transformative potential of ob-

---

4 Fanon, *Black Skin, White Masks*.
5 Rancière, *Dissensus*.

serving as a practice in terms of being attentive to one's surroundings. However, observation alone will not suffice since it does not necessarily lead to action or to particular kinds of intervention.[6] Joaquín acted because he saw himself as responsible and his intervention was tied to his duty as a Brown Beret.

Furthermore, this scene shows both the anticipation of one another's reactions on the part of Joaquín and the police, as well as their mutual awareness of observing the park and the people in it, including tourists and the local Chicanx community. Joaquín seems to have been aware of this constellation. He also realized the urgency of the situation: since the police appeared intent on assault, his intervention could not be delayed, and he had only a small window of time to act before it seemed 'too late'. Joaquín feared excessive and unjustified police intervention in a space that was highly symbolically charged. In addition, the situation was made more poignant by the fact that the police also feared a threatening reaction on the part of Joaquín, whom they classified as a potential threat because of his appearance and behavior. This constellation added to a sense of urgency, a subjective, intensive time experience of precisely this moment as key moment, which further promoted the impulse to act – because harm could only be prevented by immediate intervention. This presupposes that Joaquín was skilled at spotting potential dangers and precisely recognizing the right moment to act. He had to be able to identify the corresponding patterns and signs of his social environment and interpret them accordingly.

Thus, Joaquín's assumptions were based on specific forms of knowledge that were not always fully explicable but, in the sense of tacit knowledge, could not be precisely determined or ideologically derived. Rather, this intuitive, subtle knowledge of action is guided by implicit certainties.[7] In this specific case, this knowledge includes stereotyping and the anticipation of discriminatory actions. More broadly, many Chicanxs possess certain skills that enable them to read a social situation and assess it as potentially threatening. As a form of situated and contextual knowledge,[8] it is linked to their lifeworld and enables them to react accordingly. This knowledge in turn promotes and intensifies their alertness in everyday life, which further produces additional knowledge. In this way, watchfulness is rooted in the experience of coloniality, but can also help to challenge and overcome this coloniality.[9]

While the park, from the perspective of some of our interlocutors, needs to be safeguarded from potential threats, the park is also a public, open-access space be-

---

6 Cf. Dürr, Beobachter:in.
7 Cf. Collins, *Tacit and Explicit Knowledge*. See also Grasseni, Skilled visions.
8 Haraway, Situated Knowledges.
9 Cf. Whittaker/Dürr, Vigilance, Knowledge, and De/Colonization.

neath the Coronado Bridge. This openness complicates the guarding of the park but is in line with the non-essentializing self-understanding of Chicanxs, which emphasizes inclusion, plurality and diversity and wants to offer points of connection for like-minded people. Thus, an a-priori exclusion should be avoided. Using public space together yet separately has been framed by Vered Amit in her study about a park in Montreal as "attentive co-presence,"[10] which means taking into account one's own plans and intentions regarding the park in relation to those of others. This "watchful indifference"[11] means being observant of those visiting the park, but without necessarily engaging or interacting with one another. This practice makes it possible to use the park side by side and jointly at once – even if this requires at least some degree of consensus on how one should behave in the park.

Thus, guarding Chicano Park is not an easy task as it entails recognizing belonging and allowing for heterogeneity, but also identifying looming dangers. This is particularly important as harm can occur at any time and take different forms, be it vandalism to the murals or activities not authorized by the *Chicano Park Steering Committee* (CPSC), such as the use of the park as a backdrop for commercials, or police training exercises. The need to be watchful and to look out for signs of threat is further fueled by the growing gentrification of Barrio Logan, which encourages the intrusion of non-Chicano forces and contributes to the heterogeneity of the barrio (see chapter 5). These potential threats result in a latent watchfulness and echoes what Amit has framed as "quotidian watchfulness."[12] It refers to a state of alertness towards outsiders and potential harm, which is grounded in an intimate and caring relationship with the park. The heightened attentiveness relates to being aware of anything unusual that might pose a threat in a well-known space.[13] Quotidian watchfulness is particularly embodied by the members of the Brown Berets and the CPSC, who act as the park stewards. They are attentive to the park's use in the spirit of the Chicano movement. This is even more the case than with other Chicanx heritage sites because the park became a testimony of Chicanxs' political struggle and can be experienced as a manifest part of Aztlán, their spiritual homeland.

As we have seen in the ethnographic vignette, if suspicious behavior that is perceived as disrespectful or threatening occurs, then additional action is required to defend Aztlán. It is a matter of intervening in the right moment to avert assaults and thus move closer to the movement's goals of a future without oppression and coloniality. Thus, Joaquín's negative anticipation of police action interweaved sev-

---

10 Amit, Rethinking Anthropological Perspectives, p. 51.
11 Ibid.
12 Ibid., p. 56.
13 See also Barenboim, The specter of surveillance.

eral temporal dimensions: on the one hand, the historical experience of oppression, which is present through the many murals on the pillars of Coronado Bridge (see chapter 3), displaying vibrantly the history of dispossession and resistance of San Diegan Chicanxs, amongst other groups. On the other hand, the personally experienced, everyday practices of discrimination, racism, and exclusion. As Henrik Vigh has shown in his study on disadvantaged groups in Belfast, Northern Ireland, and Bissau, Guinea-Bissau, these experiences lead to a watchfulness based on negative potentiality. In other words, a future negative effect is expected, whereby the absent becomes present through anticipation.[14]

In the U.S. context, W.E.B. DuBois[15] has prominently elaborated on the anticipation of discriminatory acts in this autoethnographic study *The Souls of Black Folk* based on the experiences of African Americans around the turn of the last century.[16] To capture the consequences of this racialized oppression and discrimination in daily interactions, he coined the notion of *double-consciousness*, referring to internal conflict and fragmentation of the self:

> It is a peculiar sensation, this double-consciousness, this sense of always looking at one's self through the eyes of others, of measuring one's soul by the tape of a world that looks on in amused contempt and pity. One ever feels his two-ness, – an American, a Negro; two souls, two thoughts, two unreconciled strivings; two warring ideals in one dark body, whose dogged strength alone keeps it from being torn asunder.[17]

However, we argue that there are striking differences between African Americans' and Chicanxs' experiences of self and otherness in U.S. society. In contrast to African Americans, some Chicanxs see themselves as the offspring of the original inhabitants, the Mexica, seeking to gain the stewardship of their ancestral lands because of the political boundary that, since the signing of Treaty of Guadalupe Hidalgo in 1848, separates Mexico and the U.S.[18] Another difference is that while African Americans' experiences are shaped by a history of slavery and oppression, Chicanxs' reflections on history take into consideration their active role in colonial practices and resistance to these practices at once. In this vein, Chicanx watchfulness also includes an orientation towards oneself and one's own entanglement in colonialism as well as a focus on one's own community (see chapter 6). Presuma-

---

14 Vigh, Vigilance, p. 93.
15 Drawing on W.E.B. Dubois, see also Fanon, *Black Skin, White Masks*.
16 In the light of debates on gender and intersectional approaches in more recent scholarly work, the double consciousness has been expanded to a triple consciousness. See for example Arroyo, "Roots."
17 DuBois, *The Souls of Black Folk*, p. 8.
18 Price, *Weaponizing Anthropology*.

bly, this difference stems also from the fact that it is comparatively easier for Latinxs racialized as white and *mestizxs* to assimilate than for African Americans. For them, the fixation by the white gaze, as expressed by DuBois, and also by Fanon might be less strong.[19] This may even result in additional pressure for the younger Chicanx generation, who should not conform to the "gringo way," as articulated by our interlocutors, but rather continue the struggle against oppression.

In line with this, feminist Chicana author Gloria Anzaldúa describes this ambiguity in depth in her seminal book, *Borderlands/La Frontera*, while also pointing out the personal tensions that come with it and that must be endured.[20] In her view, the transgression of cultural boundaries is related to a non-conformist self-definition and leads to the formation of a *new consciousness*, which strengthens the critical reflection of the socio-political past and present of the U.S.-Mexico borderlands. In this vein, the *new consciousness* refers to an inner attitude and mindset that self-critically addresses one's own imperialistic traits, paradoxes, and ambivalences. Only through in-depth reflection on these ambivalences can they finally be overcome, and consequently, society transformed. Anzaldúa frames this reflection and work on the self as "inventory":

> Her [the mestiza's] first step is to take inventory. [...] Just what did she inherit from her ancestors? This weight on her back – which is the baggage from the Indian mother, which the baggage from the Spanish father, which the baggage from the Anglo?[21]

This specific condition is inextricably linked to the notion of *mestizaje*. The Chicanx subject position is thus defined by its heterogeneity, which in its flexibility "helps decenter the dominant subject,"[22] as it eludes a clear cultural anchoring. It simultaneously classifies in-betweenness not as an anomaly, but rather favors it as a self-description. This term is particularly suited to reflect heterogeneity as *mestizxs* and is not linked to any particular nation-state, but rather transcends national and cultural boundaries,[23] even if, as Anzaldúa's critics have pointed out, the emphasis on mestizaje erases both Indigeneity and Afro-Latinidad.[24] As Anzaldúa explains:

---

19 Fanon, *Black Skin, White Masks*, p. 82, p. 87.
20 Anzaldúa, *Borderlands/La Frontera*.
21 Ibid., p. 104.
22 Elenes, Border/Transformative Pedagogies, p. 255.
23 Alaniz/Cornish, *Viva la raza*; Amado, The "New Mestiza," the Old Mestizos; Schönwald, Ein Blick auf Chicanos; Hernández, *Coloniality of the US/Mexico Border*, p. 20.
24 Talante, Indigeneity and Blackness.

> The new mestiza copes by developing a tolerance for contradictions, a tolerance for ambiguity. She learns to be an Indian in Mexican culture, to be Mexican from an Anglo point of view. She learns to juggle cultures.[25]

However, this also means that from a Chicanx perspective, not only is society as a whole under scrutiny with regard to colonialist practices, but also one's own community. Furthermore, the self as a distinctly ambivalent subject must be closely observed as well: For Chicanxs, as well as other disadvantaged and racialized subjects (see chapter 6), see themselves not only as survivors of colonialism, but also reflect on their historical responsibility as descendants of colonizers and participants in coloniality. Their multiple cultural affiliations as well as their ambivalent relationship to the past and its legacies in the present demand a comprehensive, critical awareness of history as well as an intensive examination of their own positionality in the indissoluble web of simultaneous references to conquest, exploitation, and emancipation.

Chicanismo as an intergenerational project: intercommunity linkages and ruptures

Although Joaquín's vigilance as a budding community leader is exemplary, it is far from true for all younger Chicanxs. In the course of several decades, the Chicano movement became even more diverse through the integration of several generations with different experiences and concerns. Today, the older generation strives to keep the struggle going while also raising awareness among the younger generation to fight against inequality and injustice. The messages of the Chicano movement of the 1970s need to be kept current so that they remain effective across generations and so that individuals stay alert, that is, continue to be trucha, as our interlocutors expressed it in their own words: "Trucha, trucha, estamos en la lucha," a Chicanx scholar told Eveline (see also chapter 3). However, the journalist and son of Mexican immigrants, Gustavo Arellano, argues that the movement's programmatic founding documents, such as the Plan de Santa Barbara and the Plan Espiritual de Aztlán, are relics from the 1960s that are hardly read anymore. In addition, some members of the younger generation have adapted to Anglo society – sometimes unconsciously – more than the older generation, to the point of assimilation. While the older members of the Chicano movement were once idealistic and radical, some younger Chicanxs seem to be much more pragmatic in their

---

25 Anzaldúa, *Borderlands/La Frontera*, p. 101.

orientation.²⁶ For example, it is noteworthy that although the current membership of San Diego State University's student Chicano movement M.E.Ch.A. (Movimiento Estudiantil Chicanx de Aztlán) frequently reflects critically on the nexus of coloniality and capitalism, it nonetheless regularly celebrates a "MEChista of the Week" – similarly to how many American stores announce "Employees of the Week."

Nevertheless, to this day, many younger Chicanxs are attentive to potential forms of disadvantage, often racially based, and are alert to both discrimination and other abuses, such as land grabbing, displacement, or the restriction of civil rights, and are willing to intervene against these processes if necessary. As we observed, the younger generation also emphasizes political awareness in the sense of critically reflecting on themselves and others as important characteristics of their community. Moreover, the "community" seems to form an even more important reference point for their engagement and for their identification as Chicanx, compared to the struggle against inequality as central concern in the discourse of the older generation. However, the younger generation also identify social inequality and institutionalized racism as essential characteristics of their society, even if they belong to a more affluent U.S. middle class. For example, some Chicanx intellectuals are successful and socially mobile in U.S. society – because this society is not only racially discriminating, but also permeable and enabling prosperity.²⁷

It is important to note that the Chicano movement as a political, emancipatory project offered an alternative to the homogenizing ideas of the nation. As such, it also offered individuals of very different origins a framework for identification that also became widely visible in popular culture and developed considerable political weight.²⁸ Further, the Chicano Movement also found a welcome echo in the academic realm, in particular, within the framework of postcolonial studies that sharply criticized the colonialist ordering of the world. As a result, "Chicano Studies" were established at numerous universities in the United States, which, unlike other civil rights movements, also turned "Chicano" into an intellectual project that to this day pays special attention to reflecting on concepts such as border, identity, citizenship, and education. In line with decolonial approaches, many Chicanx Studies scholars research how colonial structures permeate and are normalized in many social spheres. This concern can be seen as the root of the political conscious-

---

26 Arellano, Raza Isn't Racist.
27 For a detailed discussion regarding Hispanic and Latinx as social categories, see Gutiérrez/Almaguer, *The New Latino Studies Reader*, and Sánchez, *Homeland*.
28 See for example, the wide reception and impact of Corky Gonzales's poem "I am Joaquin," published in 1965.

ness that characterizes Chicanxs to this day and is central to their self-image, challenging ideas of "purity" and ideologized "whiteness."[29]

In any case, the relevance of the Chicano movement must be kept current across generations so that obligations for watchfulness – that is, a sense of what we could call *trucha*-ness – endures. This is by no means self-evident since watchfulness cannot be maintained as a permanent condition. Rather, for trucha-ness to persist, a specific goal that goes beyond the interests of an individual is required. The relevance of this communal goal has to be generated again and again. In addition, injustice should be not only recognized, observed, and evaluated on the cognitive level, but an impulse for action is necessary in order to transform the social conditions. Thus, specific mechanisms need to be established to keep the struggle meaningful and to counteract the waning of commitment to the goals of the Chicano movement. This is done by conceptualizing the past not as past, but as continuing, and discarding the boundaries between different temporalities. It is necessary to act presently to move closer to a more just society in the future. The struggle is past, present, and future at once. In this sense, temporality is used to challenge and ultimately transform the existing social order. Through this chronopolitical act, Chicanxs strive to produce a different, more just and decolonial reality. At the same time, this mobilization and politicization of time is intended to create an intergenerational bond, but one that is often articulated in the common vocabulary of past, present, and future, as expressed in the words by Chicana activist Rita Sánchez:

> If our stories do not reach the next generation, our job is not complete. The greatest gift we can give to the future is our connection to one another. That sense of unity has always been the byword of our community. It is what makes us strong. That we are Hispana, Mexicana, Indígena can never be denied. Without this great understanding, we are lost. With it, we can go forward ready to build the future.[30]

Thus, Chicanxs create their own time and space constellations, whereby references from the geopolitical history of the U.S.-Mexico border region intertwine with every-day experiences of discrimination on the micro-level of daily interactions.

---

[29] There are overlaps with the concept of "la raza," which was developed by José Vasconcelos in the course of nation-building in Mexico. It refers to the idea of a "raza cósmica" as a blueprint for a Mexican nation that includes the Indigenous heritage. However, this approach is not free of racism; rather, the "bad" traits of a "race" should be dissolved through amalgamation. (Cf. Palacios, Multicultural Vasconcelos). Nevertheless, it is important to note that many of our interlocutors use the term Raza to connect themselves with other similarly racialized people in the Americas and to overcome national borders, foregrounding the struggle they share (see also chapter 1).

[30] Sánchez, Learning from the Past, p. 251.

Temporality is central here in several respects: the past is reflected as a power-bound construct and the present represents a kind of transitional period in which decisions of far-reaching significance are made in order to establish a self-determined future. Therefore, it is crucial to be always *trucha*, to be committed to the community and to not lose sight of the movement's goals. This call is particularly powerful when it is articulated in Aztlán – the Chicanx spiritual homeland.

Chicano Park as Aztlán: Challenging colonial understandings of time and space

"By taking Chicano Park, the 'myth' of Aztlán metamorphosed to reality," proclaimed Marco Anguiano in 2000 on the occasion of the annual commemoration of the park's founding.[31] He further states that

> Aztlán – the southwestern United States was the ancestral land of the Aztecs. These ancient people migrated to the Valley of Mexico and founded an empire whose capital was Tenochtitlan, now Mexico City. By claiming Chicano Park, the descendants of the Aztecs the Chicano Mexicano people begin a project of historical reclamation. We have returned to Aztlán – our home. […]. Chicano Park has provided us with the freedom to practice and express our ideas, our culture and our traditions. In short, the struggle for Chicano Park has become symbolic of our Raza's struggle for self-determination, our right to Aztlán and who we are as an indigenous people.

As a member of the CPSC, which manages the park in the spirit of the founders, Anguiano makes it clear that Aztlán is not to be dismissed as an imagination of a mythical time but is existent in this specific place here and now. He refers to the emblematic character of the park, which has established itself as a landmark of the Chicanx community in San Diego and makes it visible in the context of the city. At the same time, the social history of Chicanxs is condensed in the park as a kind of micro-cosmos, from the taking of land after the Mexican-American War to racialized discrimination and criminalization in the present. As outlined in chapter 3, the park is also evidence of successful resistance, heroism, and victory over social exclusion and racialized inequalities.

At the First National Chicano Liberation Youth Conference in Denver in 1969, the participants adapted a manifesto called *El Plan Espiritual de Aztlán*, written by the San Diego-born poet Alurista. The term Aztlán refers to the presumed place of origin of the Mexíca (including the Aztecs), who migrated from the "North" to the high basin of present-day Mexico. They settled there in the fourteenth century and

---

31 CPSC website https://chicano-park.com/cpscbattleof.html, see also Ibarra, El Campo, p. 132.

Figure 9: *Somos un Pueblo sin Fronteras.* Banner in the Centro Cultural de la Raza.

pursued a policy of expansion based on warfare and the collection of tribute, which only came to a halt when they were conquered by the Spanish-Tlaxcaltec army in the sixteenth century. Apart from this pre-colonial connotation, Aztlán refers to parts of what is now the southwestern United States, which until the 1848 Treaty of Guadalupe Hidalgo belonged to Mexico and formerly to the Viceroyalty of New Spain (see chapter 1). Thus, Aztlán is associated with the southwestern United States and is often considered a spiritual birthplace of Chicanxs. However, as a dynamic concept, the meaning of Aztlán has been continuously transformed over time and is in flux to this day.[32]

Challenging the prevailing historiography, the term Aztlán has, in addition to the cartographic dimension, a strong spiritual component and stands for home and security, but also for energy and resilience.[33] In addition, Aztlán spans different times and spaces, eluding the dominant notions of definable time periods. Rather, Aztlán develops its own chronology by pointing back to a better past, while at the same time reaching into the future. Aztlán reaches from pre-colonial cultures to the present, challenging current demarcations of the nation-state and political borders. In this sense, the decolonial debate concerning Aztlán revolves less around the question of opening or closing borders, but rather around a general questioning of their being taken for granted.[34] The slogan "somos un pueblo sin fronteras" (Fig. 9), as seen, for example, on a banner in San Diego's Centro Cul-

---

[32] Cf. Cooper Alarcón, *The Aztec Palimpsest*, pp. 22–25; Watts, Aztlán as a Palimpsest; De La Torre/Gutiérrez Zúñiga, Chicano Spirituality.
[33] De León/Griswold del Castillo, *North to Aztlán*; Hidalgo, *Revelation in Aztlán Scriptures*.
[34] Hernández, *Coloniality of the US/Mexico Border*, p. 187.

tural de la Raza, expresses that cartographic as well as epistemic disobedience is the order of the day.[35] Thus, Aztlán can be understood as an inclusive concept, both for those individuals who claim their families to have never left the area as well as those whose families have migrated from south of the border.

Significant for Chicanxs' self-understanding is the broader context of this borderland and its socio-cultural texture, which was produced by an ambivalent and extremely changeable history. Under the auspices of civilization and missionary work, Spanish colonizers incorporated the territory and its population into contemporary European history. This process was accelerated after the Mexican-American War in the mid-nineteenth century, as a result of which large parts of the Mexican territory fell to the United States. The subsequent immigration of English-speaking settlers, driven by the expansion of the railroad network and the gold rush, and ideologically underpinned by the imaginaries of powerful concepts such as *frontier* and *manifest destiny*, dramatically transformed the population. The settlers conjured up the dawn of a "new era."[36] In their perspective, history takes its course inexorably according to divine will, with humans as the executing organ of an irreversible, social Darwinian transformation in a "forward" direction on the timeline of civilization. Here, too, the political dimension of time becomes apparent, however, these processes, which oscillate between civilization and savagery, proceed at different speeds, which is why the development of society takes place in a kind of continuum from primitive to progressive or modern. This includes the positioning of cultures on temporal axes that seem to develop at different speeds and are thus classified as backward, traditional, or modern.[37]

It is in these dominant understandings of time, in its seemingly "natural" progression, including its division into periods, that what Aníbal Quijano has called the "coloniality of power,"[38] becomes manifest. These hierarchizing temporalities are grounded in a specific understanding of time, which form the bases for decisions about (chrono)political actions. Time is thus a political phenomenon and can become a source of legitimation for colonialist practice.[39] Thus, decolonial efforts oppose universalistic assumptions of time and rather point to its heterogeneous and power-related character.[40]

---

35 Ibid.
36 Mills, The Chronopolitics of Racial Time, p. 309. See also Veracini, *Settler Colonialism*.
37 Cf. Fabian, *Time and the Other*; Meinhof, Die Kolonialität der Moderne.
38 Quijano, Colonialidad del poder; Quijano, Coloniality of Power; Hernández, *Coloniality of the US/Mexico Border*.
39 Wallis, Chronopolitics; Klinke, Chronopolitics.
40 Chakrabarty, *Provincializing Europe*; Wilk, Colonial Time and TV Time.

In Aztlán, Chicanxs escape the hegemonic fixation of time and space. They instead establish their own time-space-constellation by merging different temporalities, whereby the path to a better future leads through a past that is not the past. Watchfulness is an essential tool to move from a disturbing present to a better future. In this sense, time functions less as a structuring, sequential ordering principle, but rather time, in its interweaving of past, present, and future, becomes effective both on the macro-level of hegemonic political history and on the micro-level of everyday experience: Chicanx resistance to discrimination, as evidenced by Joaquín's intervention, is permeated by its own temporality, which undermines powerful structures.

In order to grasp the texture of Aztlán, several concepts have been proposed. Because of Aztlán's constant flux and its existence in the present, terms such as "myth," "history," or "utopia" fall short – Anzaldúa also describes Aztlán as a home of border crossers and *mestizaje* that eludes temporal-spatial fixations.[41] In a similar vein, Cooper Alacrón frames Aztlán as a palimpsest that can accommodate both different times and the diversity of Chicanx identities.[42] Aztlán continues to be in transformation, while traces of the various elements remain visible. However, a palimpsest metaphor implies layers of history and diverse identities, rather than their entanglement and mutual penetration, which ultimately shapes an endless process of blurring and becoming. In any case, a key feature of Aztlán is that it does not prioritize any particular epoch but allows for diversity and the overlapping and merging of times, cultures, and also of contrasting narratives.

Aztlán became a key concept of the Chicano movement and served as a political agenda for its members.[43] However, over the course of time, the Chicano movement increasingly diversified and gave rise not only to moderate forces but also to separatist and nationalist tendencies that claimed parts of the southwest as their own territory. Framed as a re-conquista, their aim is to create an independent Chicano nation that claims postcolonial citizenship with its own national territory located in the U.S.-Southwest. However, since these projects are akin to the colonialist model and tend to invoke a static, monolithic, and ahistorical model of time and space,[44] other branches of the Chicano movement decidedly distance themselves from this approach and instead plead for an alternative conception of society, which is less mimicking but rather overcoming the current social structure.[45] In this sense, decolonial approaches seek to create alternative social orders that do

---

41 Anzaldúa, *Borderlands/La Frontera*. Cf. also Watts, Aztlán as a Palimpsest.
42 Cooper Alacrón, *The Aztec Palimpsest*.
43 Gómez-Quiñones/Vásquez, *Making Aztlán*; Anaya et al., *Aztlán*.
44 Cooper Alcarón, *The Aztec Palimpsest*, p. 7; Watts, Aztlán as a Palimpsest, p. 311.
45 Cf. Hernández, *Coloniality of the US/Mexico Border*.

not merely envisage a reversal of power structures, but rather argue for a pluralization of societies and bodies of knowledge that do not have to be oriented towards European values. They challenge the uncritical reproduction of epistemological assumptions and promote co-production and interweaving of knowledge, and discourses that run contrary to one another. This is in line with what is framed as *border thinking* or *border epistemologies*, and is prominent not only among Chicanxs, but also among numerous Latinx intellectuals who call for a paradigm shift in the course of the decolonization of knowledge.[46] With regard to Aztlán, this means that it is a controversial concept and not only acts as unifying and assembling Chicanxs, but also contributes to their division.[47] Further, it is interesting to note that despite its significance for, and popularity in the Chicano movement, Aztlán remained an almost irrelevant concept in Mexico. This seems to be the case as well for some Mexican migrants living in San Diego and visiting Chicano Park.[48] For example, a Mexican American woman told Catherine that she had been living in San Diego for twenty years before she first visited Chicano Park with a white friend, after which she expressed pride in the representation of what she interpreted as "Mexican," not Chicanx, heritage.

It is also important to note that Aztlán as a term or discourse does not necessarily figure prominently in daily Chicanx interactions. Flor, who is a volunteer in her mid-thirties at the Centro Cultural de La Raza, which labels itself as "Made in Aztlán," stresses less the spatial than the mental facet of this concept. In a Facebook message, she suggested that the term

> stands more for a kind of anti-colonial consciousness and way of living than just a concrete place. (...) I see it [as] a term/concept/story that people used at a certain point to have some reclaiming over space, very vague definition by the way. But I don't use the word, the word at [centro] right now is not really [used].[49]

In order to understand the significance of Chicano Park as Aztlán in Barrio Logan, it is important to situate the park in the history of the Chicanx community more broadly. As we discussed in chapter 3, an alliance of Southern Californian Chicanx organizations, together with the newly formed CPSC, succeeded in changing the city administration's plans, so that the park could finally be dedicated on April

---

46 See, among others, Walsh, "Other" Knowledges, "Other" Critique; Mignolo/Walsh, *On Decoloniality*; Mignolo/Tlostanova, Theorizing from the Borders; Grosfoguel/Hernández/Velásquez, *Decolonizing the westernized university*.
47 Cooper Alcarón, *The Aztec Palimpsest*, p. 10.
48 Kühne/Schönwald, *Eigenlogiken*, p. 290.
49 Flor, 25 November 2020.

Figure 10: Map of the United States showing the area pertaining to Aztlán.

22, 1971.⁵⁰ However – and this is a central aspect – based on their experience, Chicanxs fear that land can be taken from them again, which is why it is all the more important to preserve, guard, and continuously mark the park as Aztlán in order to prevent it from being taken over by hegemonic forces. This is achieved, on the one hand, through the architecture, such as the *kiosko* resembling a Mesoamerican structure and the striking murals painted on the pillars of the bridge that depict the struggle from a Chicanx perspective. Following on from the Mexican muralists, the murals pursue educational and revolutionary goals and refer to the movement's genealogy and heroic figures, deliberately picking up motifs from different times, cultures and events and directing their messages both to Chicanxs themselves and to outsiders.⁵¹ The murals serve to heighten political awareness and provide a continuous reminder to watch out against encroachment, as they depict

---

50 Ortiz, "¡Sí, Se Puede!"; Le Texier, The Struggle against Gentrification, p. 228.
51 Cf. McCaughan, The Border Crossed Us.

land grabbing and overreaching as inherent features of Chicanx history and experience (see chapter 3). The CPSC symbol features a map showing the territory of the United States with the territory of the states of California, Nevada, Arizona, New Mexico, Texas, Utah, Colorado, and Wyoming marked in red as "Aztlán." A hand from the direction of Mexico grasps the southern border of California, representing at the same time Baja California (Fig. 10).

Locating Aztlán in Chicano Park created a space of possibility for the development of a self-determined, performative, and popular Chicanx culture, which is a strong feature of the Chicano movement to this day. Performative and ceremonial practices that are carried out with a certain regularity such as danza Azteca, and musical events, political actions and art projects that are situated in the ideological context of Chicanx culture, bring Aztlán to life in the park. Of particular importance are the annual commemorations of the park's founding. During these celebrations, speakers deliver testimonies and make appeals, calling for people to remain watchful and continue to commit to the goals of the movement. As becomes evident in the next section, these occasions stimulate action and generate Chicanx political subjects. In addition, these events create and maintain a bond between the generations by the confluence of different strands of time.

Performing the struggle across time and space

> But the number one thing is to always organize, we can't walk around the park, we can't enjoy the park, if we're always being harassed. And that's why we always have to stay organized, we always have to be committed to La Causa, the movement and our people. […] And always stay calm, because a calm Chicano is a strong Chicano. […] Never lose it, cause we are stronger than that, even though things are hectic right now, times are tough, we are stronger together. United we stand, divided we fall. So that's my message to you out there. I could keep on going on with the history of the Park, you know Chicano Park is a historical landmark, not just because it's ancestral land, but because it's always been historical. To the Chicano nation, to the Chicano people and to our Chicano leaders. Many of the leaders have spoken at this park, so take in mind, when you step on that soil, many leaders, many generations, great generations have stepped on that same soil. And you're part of that history. So, every time you come to the park, enjoy it! Take in the culture, take in the spirituality, embrace it, honor it, respect it, preserve it! But also get involved! Because the movement has never died, we're still around! La Causa is still around, the issues are still going. I'm pretty sure you're done and tired of being harassed […].[52]

---

[52] Speech by a Comandante of the Brown Berets on 25 April 2020. This speech was streamed live on social media.

As part of the live-streamed celebration of the park's 50th anniversary in 2020, a veteran comandante of the Brown Berets, gave this emphatic speech. He stood uniformed and with folded hands at a podium, accompanied by two of the youngest members wearing only their brown berets but not their full uniforms. He described the park, which he called the "heart of Aztlán," as the ideological-spiritual breeding ground of Chicanx consciousness. In his urgent appeal, he repeatedly invoked the commitment to continue to adhere to "La Causa" and to remain faithful to the movement in order to ward off harassment, to join the succession of the previous generation and thereby become part of the community and its history ("you're part of that history"). As a manifestation of Aztlán, it must be preserved and defended as an asset that has been lost to Chicanxs throughout their history – both in terms of their territory on the macro-level and of their community space on the micro-level. The negative experiences of the past are recalled, and likewise the perception of the present is equally critical. For this very reason, it is important to remain united and vigilant, because the concerns of the movement are still relevant ("La Causa is still around, the issues are still going").

The alternation of time-periodizations in the comandante's address, as well as the simultaneous interweaving of different temporal structures, which are not sequential but present at the same time, intensifies the appeal for vigilance. This perception of time is epitomized in Aztlán, which is constituted as a conglomerate of different times interlaced with each other. Simultaneously, the interplay of personal negative experiences and skepticism about the future makes constant watchfulness imperative. If the duty of vigilance is not fulfilled, the community threatens to break up, increasing the pressure on the individual and the call "to be committed to La Causa" gains weight. The precise goal of this movement, however, remains largely implicit and vague, framed as a struggle for "justice" and a better quality of life – except for more concrete articulations in key texts from the movement's founding period. This vagueness is necessary, as among Brown Berets and the Chicano movement more broadly, there can be strong disagreements over which strategies and what degree of militancy should be employed exactly.

The notion of "our people" and the call to be united, which is stressed in the Comandante's speech, can be read against the backdrop of Chicanx understandings of community. As shown above, it refers more to individuals with similar experiences of coloniality, which, however, does not exclude differences.[53] Rather, differ-

---

53 Nonetheless, as Kühne, Schönwald and Jenal (Bottom-up memorial landscapes) show, this does not mean that all Mexican-Americans who come to Chicano Park necessarily feel a sense of belonging there.

ences are key in Chicanxs' self-understanding, which makes a clear contouring of membership characteristics almost impossible.[54]

During anniversary commemorations and other important events in Chicano Park, the comandante expounded his view of society and claimed interpretive authority over political processes. It remains to be seen whether he could succeed in a similar way outside Chicano Park/Aztlán. In the time-window of solemnity and in this space of possibility, however, he can gracefully unfold his interpretive power over specific concerns of the community. His performance reveals the importance of speech and making oneself heard by disadvantaged actors who, in their own words, articulate alternative perspectives to common interpretations of the world and social order that otherwise receive little attention.[55] Especially in (de)colonial contexts, the clout of one's own speech has an empowering effect on the subject formation that Jacques Rancière[56] sees as a political process in which the ability to speak is accompanied by a turning away from existing structures and, at the same time, the making visible of disadvantaged collectives. Fanon[57] also points out in his studies on racism and colonialism that the colonial subject first has to escape from the fixation hegemonically assigned to it.[58] The comandante took this step in the celebration in a similar way to Joaquín in Chicano Park. He also identified himself through his uniform as a Brown Beret and thus as an organizer and guardian of the community, continuing the line of leaders and role models across generations who have fought for "la Causa." In doing so, he lent authority and authenticity to his speech as he referred not only to previous generations depicted on the murals in the park, but also to his own experiences as a Brown Beret.

Speaking for himself is an act of self-empowerment as a knowing subject.[59] In this case, knowledge is based on the speaker's own experiences as well as that of his predecessors and that of the addressed audience. The Brown Beret comandante drew on a shared body of knowledge the collective considers as "truth" and from which normative obligations are derived. Truth, however, is produced within specific power contexts and politics. Drawing on Fanon, Daniele Lorenzini and Martina Tazzioli elaborate on the production of truth in colonial contexts and argue that "the colonized subject is seen as a subject *incapable of truth*."[60] These subjects

---

54 Pisarz-Ramirez, *MexAmerica*, p. 214.
55 Spivak, Can the Subaltern Speak?
56 Rancière, *Das Unvernehmen*.
57 Fanon, *Black Skin, White Masks*.
58 Cf. also Lorenzini/Tazzioli, Confessional Subjects, p. 176.
59 Cf. Skinner, Foucault, Subjectivity and Ethics.
60 Lorenzini/Tazzioli, Confessional Subjects, p. 76 (italics in original).

therefore stand in contrast to Foucault's conception of the "Western" subject, which is seen as able and willing to speak the truth.[61] In Chicano Park, however, this fixation and imposition of a deceitful image of subjected subjects is challenged and reversed. By claiming the truth, the comandante subverts the asymmetry of power-relations. His address refers to a Chicanx body of knowledge and essentially repeats it without adding anything fundamentally new. Rather, most of those present are likely to already be familiar with what has been said. However, it gains in topicality and significance by the fact that it is not conveyed in an abstract way but stems from the life experience of a person who places himself in the service of "la Causa" in an exemplary manner and thus articulates a kind of self-testimony.[62] In this way, he produces discourses that are essential in subject making and ultimately in the recognition of these subjects as telling the truth.

The reference to several role models increases the obligation for the following generation to continue the struggle. The goal of a just society has not yet been achieved, the concerns of the movement are still acute, which is why the struggle must be continued ("issues are still ongoing"). With this argument, the comandante points beyond the Chicanx collective and addresses the broader social context that continues to necessitate this very movement. This concern gains poignancy by referring not only to the past, but to the ongoing need for action in the present because of the grievances that continue to exist. To transform society and to achieve the movement's goals, the audience, especially the younger generation, is asked to become actively involved in the struggle – without specifying this act in more detail ("get involved").

In order to maintain the collective nature of the Chicano movement, the comandante addressed the commitment of each individual as essential. Social change can occur through a commitment to both commonality and communality. Emphasis is placed on ongoing structural disadvantage, spatial displacement, and a pressure to assimilate to Anglo-American hegemony. This makes it all the more important to stress one's own values and obligations towards Chicanismo. In this way, in addition to the evocation of community, boundaries are drawn towards an "outside" of Chicanx sociality,[63] which has a unifying effect towards the "inside."

In the Comandante's speech, he refers not only to current members but also to individuals from the past who serve as role models. Thus, he includes recently deceased people, who are integrated into the "struggle" after their death. Those who had committed themselves to the movement during their lifetime are also consid-

---

61 Foucault, *Discipline and Punish*; *On the Government of the Living*; see also Lorenzini/Tazzioli, Confessional Subjects, p. 76, p. 81.
62 Cf. Hartmann/Jancke, Roupens Erinnerungen.
63 See among others, Butler, *Kritik der ethischen Gewalt*.

ered *presente* after their death – as still present and as visibly belonging to the community. Their presence is evoked during multi-day funeral services, as was the case with a Brown Beret elder, who was esteemed in the community as the "Honorary Mother of the Brown Berets" and who was with the "struggle" from its earliest days. After a two-day memorial program was held in her honor immediately following her death in 2020, a flag-raising ceremony was held in June 2021 to mark her birthday. The program again included speeches, music, and drinks. The public Facebook invitation showed her as a smiling older woman in Brown Beret uniform and wearing a necklace of colorful beads in the style of Native American jewelry. The post read, "Join us as we celebrate a true chicana revolutionary that dedicated her life to the struggle for self-determination and national liberation. ¡[...] PRESENTE!" There is no finality to the struggle and even as she passed away, she continues to be present as an exemplary community member.

In this Chicanx time-space context, also those who lost their lives because of racist structures remain present and visible. A miniature Mexican cemetery-style memorial under one of the bridge pillars in Chicano Park commemorates Mexican-born people who died when the car of a drunken Anglo driver plunged off the bridge into the crowd (see chapter 3). The memorial, which is maintained and redecorated several times a year, depicts the deceased in framed pictures, as well as the lasting affection for them and indignation of the mourning community over their deaths. Barrio Logan residents interpret the accident not as a tragic coincidence, but as a predictable consequence of racist, negligent urban planning and evidence of existing discrimination. These discourses continue to mark the struggle as significant across generations – and establish an involvement that extends beyond death. Therefore, the life cycle or presence of Chicanxs does not end with death – it is the deceased heroes of the movement and survivors of oppression who continue to be *presente* as role models and transcend time in its finitude. It is this interweaving of different temporal strands that is significant for Chicanxs' understanding of the world.

We have shown in this chapter that various forms of Chicanx watchfulness in San Diego are shaped by the specific socio-political conditions of the U.S. Southwest. A striking feature of these borderlands is a tension between attempts to sharply classify and separate the population and the simultaneous impossibility of this endeavor, which is permanently undermined in everyday life. Chicanxs embody these contradictions and often claim citizenship rights in accordance with the U.S. constitution yet are continuously exposed to the suspicion of actually living on the "wrong" side of the border.[64] Significant for these processes is a power-bound

---

64 Flores-González, *Citizens but not Americans*.

geo- and chronopolitical division of the world that aims to fix identities in space and time and thereby attempting to plausibilize differences. These practices are fueled by media threat scenarios and racialized discourses of belonging in society as a whole, accompanied by a material infrastructure and technology of surveillance that is paired with imaginaries of impending danger from the South.[65]

Chicanxs elude this fixation by drawing on Aztlán with its own time-space dynamics linking different temporalities, whereby the path to a better future leads through a past that is not past. Watchfulness is a tool helping to move from a disturbing present to a better future. In this vein, time functions less as a structuring order in the sense of a linear succession, but rather time becomes effective in its interweaving of past, present, and future both on the macro-level of dominant political historiography and on the micro-level of everyday experience: Chicanx resistance to discrimination is permeated by its own temporality, which undermines powerful established orders. It is this interweaving of different temporal strands that is significant for the actors' understanding of the world. This becomes particularly evident when those who join the Brown Berets or the CPSC remain members for life. Even after death, members, transfigured into heroes and victims, continue to be *presente* in the community as role models of the Chicano movement. In this way, they transcend the boundaries of Anglo-American time.

---

65 Chavez, *The Latino Threat*.

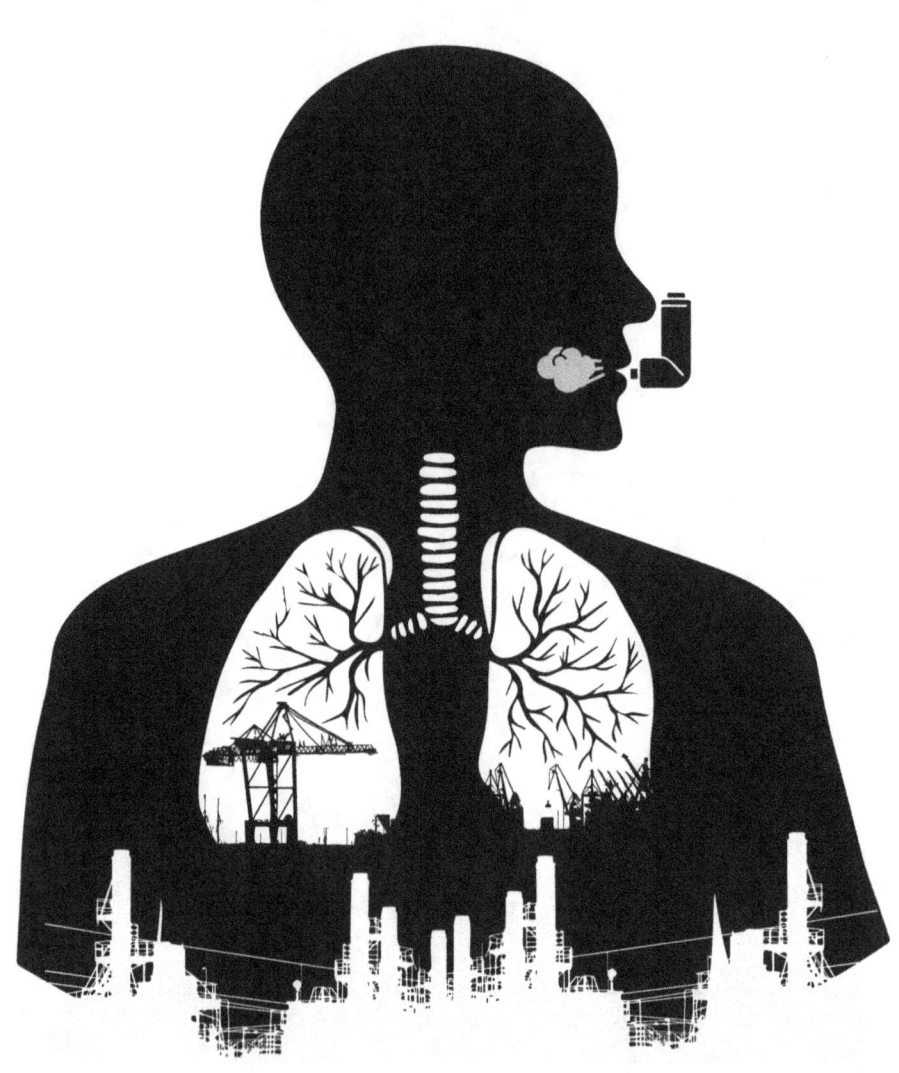

Chapter 5
"Why us?": Making Environmental and Health Threats Visible

"WHY US": The letters are bold and white, engulfed in flames, hovering over a grinning skeleton man. He is wearing a black tailcoat with a big white dollar sign on the back and a top hat with a skull-and-crossbones symbol on it, a toxic figure representing major economic interests: the National Steel and Shipbuilding Company (NASSCO) on 32nd Street, Ryan Aeronautical (who built Teledyne Ryan reconnaissance unmanned aerial vehicles) and the Naval Amphibious Base on Coronado Island. This sinister figure is turning the handle of some industrial machinery to pour his toxic waste down a pipe below him, spewing it out into San Diego Bay under Coronado Bridge, producing the word TOXICS. As a humorous touch, the muralist added "(ASCO)" (disgust) under NASSCO, creatively juxtaposing the two words in a defiant rhyme.

Figure 11: *WHY US*, mural, Chicano Park.

∂ Open Access. © 2023 the author(s), published by De Gruyter. [CC BY] This work is licensed under the Creative Commons Attribution 4.0 International License. https://doi.org/10.1515/9783110985573-009

The simple, stark imagery of this nightmarish mural alerted us to otherwise invisible threats when we walked around Chicano Park as a group in September 2021. Like the many other murals in the park, which connect colonial history with the coloniality experienced by the present generation, this mural portrays the contemporary experience of pollution in Barrio Logan as being linked to the military-industrial-complex at the U.S.-Mexico border.[1] It also draws attention to the community's resistance to it – their *survivance* in the face of coloniality. According to Native Studies scholar Gerald Vizenor, "Survivance is not just survival but also resistance, not heroic or tragic, but the tease of tradition, and my sense of survivance outwits dominance and victimry."[2] In her 2021 zine, the San Diegan Chicana writer-activist Eréndira Jiménez Esquinca expresses a similar notion, predicated on "non-binary thinking" and "choosing the hard parts of relationship (honesty, challenge, agitation, love)":

> Suffering and joy are bound to each other [...] I've thought about how to create communities of honesty and self-awareness and flourishing. I've thought about how folks on the margins have to fight to maintain mental health, equitable learning/work environments, proper compensation in systems that weren't built for them. [...] I believe it's possible because there are [...] too many ancestors that have said a new way, a new life, a new world is possible [...][3]

For Jonathan, who had just spent several years living and thriving in the coastal town of St. Andrews in Scotland, a place known for its quiet, quaint remoteness, frequent rain, and fresh sea breezes, walking under the Coronado Bridge, ever roaring with traffic, towards César Chávez Park was an assault on his senses. While the rest of the team was fairly oblivious to it, Jonathan could feel the pollution in the air, which was clogging up his sinuses and forcing him to cover his face with a mask. He wondered at how local people manage living in such a polluted environment.

In the vein of survivance, we learned that the people of Barrio Logan live varied lives, often making space for play and humor amidst the sorrow over the ill health and death of many locals, in which the pollution is a major factor. Barrio Logan is among the top 5% neighborhoods for air pollution in California.[4] Many people in Barrio Logan breathe a poisonous mix of fine dust, diesel particles, and black carbon every day, all day. This means that many have developed serious health conditions that shorten their life expectancy, including respiratory illnesses, such as asthma, and other health complications, such as cancer and heart disease.

---

1 Miller, *More Than A Wall*.
2 Vizenor, *Fugitive poses*, p. 93.
3 Jiménez Esquinca, *Deabstracting Decolonization*.
4 Cavanaugh/Cabrera, Examining The Effects Of Climate Change On Barrio Logan.

The Environmental Health Coalition, a local environmental justice NGO, have conducted ongoing research within Barrio Logan on the connections between air pollution and health locally, and have campaigned for improvements in air-quality, for example, by the alteration of the neighborhood's status as a mixed-use zone. As Diane Takvorian, the director of the Environmental Health Coalition, reported, "Children in Barrio Logan and National City [also a Chicano neighborhood] have more than double the rate of asthma emergency room visits than San Diego County as a whole [...] and residents have a higher risk of developing cancer from air toxins than 93 percent of the nation."[5] Takvorian identifies diesel emissions as the main pollutant, which are particularly high in Barrio Logan as a result of its proximity to the Coronado Bridge and the I-5 highway. We discussed their imposition on the neighborhood in chapter 3.

The deadly pollution that the people of Barrio Logan live with is thus linked to having been excluded from specific city planning and environmental policy decisions. As mentioned in the introduction to this book, many Mexicans originally immigrated to Logan Heights to work in various trades, including as laborers, cannery workers, welders, pipefitters, and longshoremen. After World War II, Barrio Logan was designated as a mixed-use zone, which means that heavy industry polluting industries, junkyards, metal plating shops, were built next to residential buildings. We suggest that this amounts to a politics of death, a necropolitics, in which city planners hold "the power and capacity to dictate who may live and who must die."[6] As Sandset argues, "the necropolitics of global health inequality is driven not by a perpetual state of emergency, but by a state of chronic acceptance that some have poorer health than others,"[7] such as by setting up health care systems that are "not conducive to life but to slow death" for certain segments of the population.[8] In racialized and socioeconomically oppressed neighborhoods like Barrio Logan, people not only face greater obstacles to accessing quality health care but are also suffering under "the attritional violence of environmental pollution, often through 'violent inaction' of regulating authorities."[9] This might be described as an insidious and often overlooked "slow violence," that "occurs gradually and out of sight, [...] dispersed across time and space."[10] Moreover, not only is the production of pollution in areas where working-class People of Color live the result of ongoing coloniality, but many environmental organizations also repro-

---

5 Takvorian, Opinion: Plans to Reduce Pollution Aren't Enough for Barrio Logan.
6 Mbembe, Necropolitics, p. 11.
7 Sandset, The necropolitics of COVID-19, p. 1412.
8 Ibid.
9 Davies, Toxic Space and Time, p. 1540.
10 Nixon, *Slow Violence and the Environmentalism of the Poor*, p. 2.

duce coloniality, as several of our interlocutors highlighted. This is because they are run by Anglo-Americans who disregard that environmental and racial justice are interrelated issues and often fail to consult People of Color, despite being those most affected by environmental injustice.[11] As Jiménez Esquinca argues, conversations about "anti-racism and climate change and indigenous sovereignty and the end of capitalism"[12] are interconnected, hard, and necessary.

Calling on each other to be watchful is a key way in which Chicanxs seek to counter the (largely) invisible threat of air pollution and the racist and colonialist logics that have caused and exacerbate it. Yet watchfulness itself comes at a high cost to the neighborhood's health, as "one cannot live in a constant state of alertness, and so the chaos one feels becomes infused throughout the body. It surfaces frequently in dreams and chronic illness."[13] Perversely, those suffering are often blamed for their poor health,[14] even though medical anthropological studies have demonstrated a link between the social determinants of mortality and "the social determinants of the distribution of assaults on human dignity."[15] Underlying conditions are not equally spread throughout the population, but combined with, and are the result of, particular socio-economic factors.[16] Accordingly, People of Color are more likely to suffer from ill-health, and Women of Color even more so.[17] For instance, many of our female interlocutors reported having trauma-related chronic pain or other health issues due to having experienced gender-based violence, such as sexual abuse in their childhood.

The existing combined environmental and health injustice in Barrio Logan and similar neighborhoods was brought into sharp focus during the COVID-19 pandemic. A nation-wide study found that "Even amid COVID-19, US White mortality is likely to be less than what US Blacks have experienced every year."[18] Sandset argues that "the COVID-19 pandemic is entangled with necropolitical factors of slow violence and death that preceded the pandemic and adds to the disproportional distribution of vulnerabilities towards the risk of infection, death, and economic impoverishment."[19] The story of Diego, a member of the American Indian

---

11 Liboiron, *Pollution is Colonialism*; Méndez, *Climate Justice from the Streets*.
12 Jiménez Esquinca, *Deabstracting Decolonization*.
13 Green, Fear as a way of life, p. 231.
14 Farmer, *Pathologies of Power*, p. 49.
15 Ibid., p. 19.
16 Wood/Harris/Maglalang, A Call to Healing, p. 85.
17 Ibid., p. 87.
18 Wrigley-Field, US racial inequality may be as deadly as COVID-19.
19 Sandset, The necropolitics of COVID-19, p. 1412.

Movement with Indigenous Mexican ancestry in his early thirties, who had lived in the area since his childhood, may help to illuminate some of these entanglements.

Diego recalled that he had not taken the threat of the virus seriously until he himself became seriously ill with it. He had a longstanding distrust of the government and the health system, which was common among People of Color in the U.S. because of discriminatory health care and being targeted by misinformation campaigns.[20] This placed him at a greater risk of contracting COVID at the shipyard, where many workers like him had refused to get vaccinated. When Catherine interviewed him at Chicano Park in September 2021, he was still limping slightly because the long-term COVID effects had worsened the stiffness from his old injuries. Diego had suffered multiple gunshot and knife wounds in his younger years, as well as having been hit by a police car on purpose, he alleged. His story showed that for some young men in Barrio Logan and other working-class Latinx and Raza neighborhoods, the violence of gangs and of the police potentially pose additional serious risks to their health and survival. At least in Diego's case, experiencing and witnessing violence since his childhood had fostered a vigilant disposition in him, which had developed into clinical-level, anxious hypervigilance. Accordingly, he chose to sit at one of the picnic tables opposite the *kiosko*, which gave him a clear view of everyone who might pose a threat. During the conversation, he displayed a heightened awareness of his surroundings, for instance by pointing out a distant bird of prey in the sky as well as a police car driving through the other side of the park. While his hypervigilance was intended to protect Diego from danger, it also adds to the chronic stress that racialized discrimination causes, so that it contributes to his health conditions. Like many other migrantized Chicanxs and Mexican Americans who are subject to racialized stress,[21] Diego lived with a weighted down body and mental health challenges.[22] At the same time, Diego seemed hopeful for the future, which the pandemic had given him time to think about. He was about to get married and was thinking of starting an organization to support local boys' education and keep them out of gangs. Apart from shining a light on specific vulnerabilities and injustices, the pandemic therefore also brought to the surface solidarities and strengths in the Chicanx, Raza and other Latinx communities of San Diego.

In this chapter, we will further illuminate the ambivalent role of watchfulness in relation to how people in Barrio Logan and neighboring Chicano neighborhoods

---

20 Vaccine hesitancy and belief in conspiracy theories was also very common among white Anglo-Americans, which some have linked to a false notion of racial superiority (Metzl, *Dying of whiteness*).
21 Raschig, Cargas Coming Down.
22 Crocker, Bodily Imprints of Suffering, p. 226.

and other San Diegans address the coloniality of healthcare, which normalizes racialized inequalities. Viewed from an intersectional perspective, Chicanxs as subjects face health inequalities on the basis of ethnicity and class, as well as other aspects of their identity, such as gender.[23] Similarly, residents of Barrio Logan suffer from the worst air pollution in San Diego because of their ethnicity- and class-based marginalization. However, the watchfulness and awareness that artists, such as the muralists who painted *WHY US?*, activists, members of the Environmental Health Coalition or the Centro Cultural de la Raza, and other educators have been spreading for the last 40 years is increasingly inspiring victims of environmental and health injustice to fight back. This struggle for better health through improved air quality became particularly urgent during the COVID-19 pandemic, as we will see in the following sections.

An autoethnographic perspective on the COVID-19 pandemic in San Diego

From the 12th of March 2020, San Diego locked down in order to limit the spread of COVID-19 infections, and by the 19th of March this was part of a California-wide lockdown, which radically limited Catherine's ability to conduct ethnographic fieldwork, forcing her to temporarily shift to an autoethnographic approach (see chapter 2). People were only supposed to leave their houses for certain essential activities, which included grocery shopping, exercise, dog-walking, and visiting pharmacies.[24] Officially, people who violated stay-at-home orders could be fined or even risked prison sentences. Such measures can be regarded either (or both) as attempts to prevent transmission of the virus or as biopolitical measures creating the conditions for new forms of surveillance and control.[25] This in turn inspired suspicion and recalcitrant impulses among many San Diegans. When Catherine went to her local Mexican supermarket in Kearney Mesa a week into lockdown, people were still entering freely without wearing any masks and a visit to Home Depot found the shop full of people. However, by early April, the supermarket was strengthening its anti-virus measures; bottles of hand disinfectant were being sold at the cash and cashiers and some customers were wearing masks, though not always keeping sufficient distance from one another. There was still a considerable amount of people out shopping. People's adherence to public health

---

23 Crenshaw, Mapping the Margins.
24 City of San Diego EXECUTIVE ORDER NO. 2020–11.
25 Calderón Gerstein, COVID-19, Ontopolitics, Necropolitics, p. 78 f.

measures demonstrated at turns an indifferent, dismissive, or vigilant attitude towards the virus and concern for the health of others around them, as trust towards the government narrative around the virus varied. Catherine's host, for example, remarked on the need to be "vigilant" to avoid infecting elders in the community, including his mother and mother-in-law.

While a cautious attitude regarding COVID-19 meant that people monitored themselves for the benefit of all, they were also called by community organizations to support vendors and restaurants as a way to "defend the hood" at a time when people were conscious that they could lose everything they have worked for very quickly. As early as April 12, *The San Diego Union-Tribune* reported on the risk that COVID-19 might accelerate gentrification in Barrio Logan, as businesses deemed "nonessential" were forced to close during the lockdown.[26] One resident was quoted as saying, "If we all begin to close and lose our spaces, eventually when we come back, people who can afford it will move in and it won't be Barrio Logan." (We will return to the topic of gentrification later in this chapter.) When a spirituality- and healing-focused shop run by self-identifying Witches (see chapter 6) began making and selling cloth masks in March 2020, they urged their followers on Instagram to wear facemasks because of Barrio Logan's high-risk population of people living with asthma and other chronic conditions. Customers also often lived with people who were at high risk and were unable to isolate themselves.

Just as much of our research moved online, many community events in Barrio Logan were now held online to protect vulnerable individuals. This necessary measure was particularly tragic for the celebration of the 50th anniversary of Chicano Park, which had been planned as a huge event to mark the achievements and continued determination of the Chicano movement. Every year, Chicano Park Day is a significant event in bringing together the community to celebrate their own history, but the 50th anniversary had been intended as the culmination of this tradition. As the organizers, the Chicano Park Steering Committee (CPSC) instead had to move the celebration online, which included live-streaming Chicano soul music on the CPSC Facebook page. In addition, topical video clips were uploaded throughout the week, such as a video of a small flag-raising ceremony in Chicano Park to mark the passing of the community's work from one generation to the next, and a cleansing ritual in which copal (incense-like resin) was burned in clay burners. Photos of artwork about the park were also posted on the page.

Other online community events included bilingual online mass services, held by Our Lady of Guadalupe church. As the weeks and months went on, the list of people affected by COVID-19 in the prayer sections of the online masses became

---

26 Lopez-Villafaña, Barrio Logan community.

longer and longer, Catherine observed. She was also attending a Raza women's healing circle (*círculo de mujeres*) online every week, which had previously been taking place in person on a bi-weekly basis (see also chapter 2). Berenice, the organizer, a Boricua PhD student of acupuncture in her late twenties, had decided to increase the frequency of the meetings in order to meet the increased need for mutual support in this time of crisis and uncertainty. Among the circle attendees, the pandemic quickly seemed to take its toll. The women and non-binary individuals often reported that their mood had become dull and low as a result of isolation or being unable to work. It was difficult, for example, entertaining children indoors all day every day, as well as social distancing from family members within the same household. However, offering each other mutual support by listening to each other's stories and commenting on these without judgment, as well as practicing meditation together, helped to alleviate some of the pandemic-related stress they were experiencing. For Catherine, too, the circle was a lifeline for her mental health at a time when she felt cut off from fieldwork opportunities and from being able to return home.

Solidarity during the pandemic was an important expression of belonging to the community itself. This is because one becomes a member of community organizations by showing up and working for them long-term alongside other members. At an online "Chicano Revolt Symposium" that Catherine watched live on Facebook, one of the Centro Cultural de la Raza's board members, Roberto Hernández, declared:

> El Centro es de quien lo trabaja (The Centro belongs to those who work it). Which is to say: We have to put in work. Everybody has to put in work. When you put in the work, you too are part of the *movimiento*. Whether you're a *veterana, veterano* or someone, you know, younger folks today. You know, get involved. You put in work, the *movimiento* belongs to you too, right. That's how we organize the Centro, that's how we see the broader *movimiento* that has not ended. You put in work, you're a part of it, you own it and we build together.[27]

Similarly, the "Raza Visions II: Cultivating Creative Spaces of Autonomy & Resistance" zine that the Centro published in July 2021 states: "'El Centro es Quien lo Trabaja' is both a commitment and an invitation in recognition that the Centro Cultural de la Raza is a space for everyone who wants to contribute not to us but to our collective good."[28] Taken out of context, this statement in the tradition of Chicanx utopianism may sound romantic.[29] However, the zine authors cite Chicana

---

27 Centro Cultural de la Raza, *The Chicano Revolt*.
28 Centro Cultural de la Raza, *Raza Visions II*, p. 23.
29 Alvarez, Free Air Purifier.

theorist Cherrie L. Moraga as requiring "head-on collisions" within the collective to access "real power."[30] The Centro's understanding of tequio, of collective work, is inspired by the Zapatista Movement, and is defined as mutual support and "a conscious collective mind that understands the root causes of capitalism and colonialism, and constantly tries to break those structures of power" towards collective flourishing.[31]

This means that the Centro has developed as a close-knit community of people who actively and committedly contribute to its development but occasionally clash over what exactly this development should entail, much in the spirit of the Moraga quote above. Catherine's observations and conversations with various Centro members suggest that younger and newer members are usually given many opportunities to participate in different forms of tequio. With respect to the Centro's programming and leadership structure, when conflicts arise, all parties are heard, but typically, elders are given more support by the collective.

Around July 2020, physical in-person events started occurring again, such as a Tianguis (market) at the Centro with a dozen stalls selling crafts, and other community events, like the Women's Circle, now switched to hybrid mode, with some people attending the meeting in the Centro's garden. Catherine began taking part in the Centro's tequio by assisting with its long-term project to transform its garden. This collective work included removing old cacti, mulching the earth, planting native plants, watering the sprouts, making seating areas, and more. In the zine, one of the garden volunteers described "connecting with Tierra Madre" as a "healing journey" away from "hating myself and hurting":

> With tiny steps I began to resist the unhealthy person I had become. [...] This spurred me to remove toxins from my health and skin care products, learning how to grow as much of my own food as possible, and reduce my waste, especially plastic. Through working and volunteering in community and school garden spaces grew my biggest revolution, I started loving my body as much as I loved the soil and plants. I stopped resisting the fear of radical change and started resisting capitalist consumerism.[32]

This quote illustrates what an anti-capitalist conscious collective mind might look like and how it might challenge the coloniality of toxicity in San Diego. Taking a holistic perspective and assuming responsibility for her choices within a structurally violent context of slow death, the volunteer linked gardening to healing her spirit from the ravages of consumerism and abusive relationships as well as heal-

---

30 Centro Cultural de la Raza, *Raza Visions II*, p. 21.
31 Ibid., pp. 21f.
32 Ibid., p. 20.

ing her body from toxic food and products, while also contributing to healing her community and the earth that sustains it. In conversation with Catherine, she added that during the pandemic, gardening had also provided some comfort at a time when even participating in the Women's Circle had felt too overwhelming in the face of bereavement. In the next section, we will turn to how San Diegans experienced the coloniality of health through the inequalities of precarity and death during the pandemic, as well as discussing the watchfulness of local activists in response to the situation.

The intersectionality of vulnerability and responses to the pandemic

In a necropolitical context of racialized environmental health inequality, an individual's vulnerability to illness and to dying from it is strongly linked to where they are positioned, both socially and geographically. It was thus no coincidence that from March 31, 2020 onwards, the Otay Mesa Detention Center, which is a prison-like complex where undocumented and unauthorized migrants are held, reported the highest concentration of COVID-19 infections at a single site in San Diego County. One male detainee reported to the Chicanx organizers of the Otay Mesa Detention Resistance (OMDR) that the detention center was a "death trap" and that people had died there. In response, they organized a protest event in front of the detention center at which they drew an explicit connection between police violence against People of Color and ICE agent violence against migrants and migrantized people.

The detention center is protected by two layers of chain-link fence topped by razor wire and is located in a remote, industrial part of San Diego County near the U.S.-Mexico border and the county's correctional facilities. Run by the for-profit corporation Core Civic, the two-story building appeared like a visually intimidating, impenetrable block in the arid landscape. At the protest, attended by 40-or-so mostly young adult Latinx protesters, one young Latina took the microphone and explained that she was a former detainee, pointing out the bitter irony that in the so-called "land of the free [...] in reality more people are incarcerated than anywhere else in the world." She added, "They treat people like they're not human beings," accusing Core Civic of taking insufficient COVID precautions. She alleged that Core Civic used window cleaning material instead of proper sanitizer. In the detention center, COVID information had only been offered in English, so the majority of detainees had been unable to understand and receive protective masks. When English speakers had protested loudly, she recalled, Core Civic's emergency team had threatened them with pepper spray and three people had been taken into "the hole." Another detainee confirmed that he had experienced bad food, filthy conditions, and inhumane treatment. Again, the crowd chanted, "FREE THEM ALL!" Finally, a singer-activist offered "poesía para la sanación" (healing poetry) because "la violencia estatal [...] Nos da toxinas en el cuerpo" (state violence ... gives us toxins in our body). Much like environmental and immigration justice activists in Los Angeles,[33] she was thus drawing a connection between the state's violence against migrantized people in the form of imprisoning and deporting them and the violence of environmental politics. Where new factories and industrial sites that contaminate the environment with toxic chemicals and waste are built is largely determined by race and class in the U.S., which some have referred to as environmental racism.[34] However, this is rarely connected to active, intentional discrimination today, instead resulting from the profit-maximizing choices of businesses and government that neglect and exploit long-standing inequalities.[35] Rather than just lament this situation, the singer's healing poetry gave hope that injustice can be overcome collectively.

For the protesters at Otay Mesa, their watchfulness during the pandemic did not emerge as a response to a new phenomenon but was an extension of the long-standing vigilance that they already employed to withstand coloniality, including environmental injustice and its effects. As José Antonio Vargas from the NGO Define American, who is himself undocumented, pointed out in May 2020, unauthorized, illegalized migrants have always faced travel restrictions, and social distancing and self-isolating is not a new behavior for them, but has long been a way of life as a result of their separation from loved ones and inability to return home to be with them in times of need,[36] as well as the constant fear of being detected by border patrol agents. While for many U.S.-Americans, the pandemic was their first experience of forced immobility and the sometimes extreme loneliness connected with it, for many unauthorized migrants the public safety measures felt like a continuation of what they were already used to.

Similarly, at the San Diegan Black Lives Matter protests that followed the killing of George Floyd by a police officer in Minneapolis, many people drew connections between experiences of police violence towards themselves and their communities and the heightened vigilance and feelings of marginality they had experienced during the pandemic. At one protest in Downtown San Diego that Catherine attended in June 2020, a young Latino activist called on those present to "overcome not only the pandemic of illness and violence but also this pandemic

---

33 Kim, *Refusing death*.
34 Checker, *Polluted Promises*.
35 Taylor, *Toxic Communities*; Sze, *Environmental Justice in a Moment of Danger*.
36 Yam, Struggles of undocumented people.

of fear and uncertainty," inciting them to "transform fear into action." Rather than just demand an end to police violence, he highlighted that law enforcement should be defunded because "Hospitals and housing need more funding" and appealed to the crowd to stay alert and continue fighting for social justice. The peaceful protest was broken up by white supremacists who tried to intimidate them by performing Nazi salutes, revving engines, making clouds of smoke and yelling racist insults.

The racism that People of Color in the U.S. face on a daily basis, and which provokes vigilant responses, intensified during the pandemic. Asian Americans bore the brunt of racist responses to Trump's rhetoric concerning the "China virus" or "Wuhan virus." Catherine's Republican host even complained to her that people weren't blaming China enough for the virus. By June 2020, there were apparently reports of over 100 hate crimes per day against Asian Americans[37] and from March 2020 to February 2021 reported hate crimes rose by 73% according to the FBI.[38] In response to feeling more threatened, Asian Americans bought 42% more firearms in the first six months of 2020 than they did in the same period the previous year, according to a survey by the National Shooting Sports Foundation (NSSF).[39] Racism against People of Color included assumptions about their greater susceptibility to the disease being based on genetics, rather than socio-economic factors. One day at home, Catherine's Korean American roommate Jay raised the question of genetic susceptibility when reading news that African Americans were more badly affected by the virus. In response, Catherine listed the many reasons why African Americans are affected: lacking (sufficient) insurance and access to medical treatment,[40] mistrusting the government, pre-existing conditions, working in jobs that do not allow for working from home, living in small, crowded homes and polluted areas, in food deserts, also encountering racist medical staff; she also pointed out that some Black men are afraid of wearing a facemask or bandana, for fear of being mistaken for gang members. In Barrio Logan and Logan Heights, many people work long hours and cannot take days off. Such precarity, which often contributes to poor mental health[41] as well, leads to COVID-19 transmission rates being higher in racialized poorer communities.[42] Jay scoffed at this. As people in Barrio Logan were aware of the lack of governmental health provision, some stepped in themselves. The Brown Berets went from door to door

---

37 Croucher/Nguyen/Rahmani, Prejudice Toward Asian Americans.
38 Venkatraman, Anti-Asian hate crimes.
39 Chan, 'I've Never Seen This Level of Fear.'
40 Artiga et al., Health Coverage by Race and Ethnicity, 2010–2019; Lukens/Sharer, Closing Medicaid Coverage Gap.
41 Jenkins, *Extraordinary Conditions*.
42 Sandset, The necropolitics of COVID-19, p. 1418.

to urge neighbors to get vaccinated at a communal vaccination event in Chicano Park. Hundreds of people attended. This echoes how the community founded their own health center through the take-over of a building in Barrio Logan in 1970.[43]

Independent of the racist implications of Jay's question, ignorance of health inequalities was widespread. Berenice, who was working as a part-time health coach to help finance her studies, expressed that people in the U.S. are very disconnected from their personal and collective health. As a result, she explained, people tend to think of health as an individual matter, like going to exercise, and remain unaware of things that contribute to collective health, like herd immunity, pollution, and so on. To address this lack of awareness, she argued that behavioral change is needed, which is difficult, and needs to be supported by coaches like herself. Moreover, people had to be watchful of their individual health because of the lack of significant attempts by governing authorities to put in measures that would improve the health of people in Barrio Logan and surrounding neighborhoods collectively. Berenice asserted that many Latinxs don't trust the government and authorities, with reason.

This lack of trust in the government seemed to be further confirmed in July 2020, in a conversation between Catherine and Aura, a Mexican and American *fronteriza* and "artivist" (artist-activist) in her late thirties from San Ysidro. They sat on park benches in front of the Centro Cultural for an hour in the cool evening breeze. Aura started off by saying that, while she considered COVID to be "a real virus," she also suspected that "it's being controlled" to harm certain populations because it had already emerged last year in China. "Why wasn't the U.S. prepared?" she asked, frowning. In her opinion, it was unnatural for the virus to be spreading in the way it was. Aura said that she wore a mask "out of respect for people" but that she did not like wearing it and wished she could just tear it off. She seemed skeptical as to how much it actually protects people and appeared more troubled by the economic effects of the virus than by the health aspect. This stemmed from the fact that she herself could not find paid work at that time and was feeling restless at night, watching funny videos, often sleeping until midday. She had always been an independent, self-reliant woman, so she did not like having to depend on her sister, who has a small baking business, but she was grateful not to be in an even more precarious position. Until a few years ago, she had been living in Tijuana with her grandmother out of convenience, due to the low cost of living. Having trouble paying rent and other bills, as well as for food, brought Aura feelings of sadness, fear, frustration, anger, and poor mental health. The overall stress

---

43 Blanco, *A Brief History of the Brown Beret National Organization.*

of the situation was such that one of her eyebrows fell out, an experience of loss that she later reworked artistically in her 2022 zine "Sin Una Ceja – without an eyebrow," where she writes: "I want my right eyebrow to grow back to its normal size and shape!"

Aura followed the news and had been attending protests for over a decade. Yet she often felt that her artivism was not enough to make a change, or even a living, wondering if she had chosen the wrong career, even though she loved it. Together with a Latinx student collective from San José State University, she had produced a zine of art and poems on their experiences of the pandemic, in which they acknowledged the intersectionality of vulnerability: for example, health conditions that are more common in non-white populations, such as diabetes and obesity, making them more vulnerable to the potentially deadly effects of the virus. While she wished she could have more influence on the situation, Aura acknowledged that, as an artist, she "can still contribute," by providing inspiration and vision to the movement. In her view, too few people want to think about politics.

Echoing Aura's frustrations about joblessness and the governments' failures in managing the pandemic, Catherine's "half-Mexican" host Christina lamented being unable to return to work because her employer, the University of San Diego, a private Catholic university, announced that they would not be allowing students to return to campus yet. She had seen the billboards on her way home from Mission Bay after going out on the boat with her husband Pete there and "it felt like the beginning of the Hunger Games, like we're being controlled." Christina and her family had previously been traumatized by the 2008 crisis, during which they fell from being well-off to struggling with their finances, which meant that Christina, who had been a full-time homemaker, had needed to get a job. The crisis also forced their son to quit university and finish his degree at a less prestigious, but cheaper, community college. Accordingly, when the COVID lockdown happened, the potential economic costs of the crisis were immediately at the forefront of their thinking, more than health concerns. However, even though she and Pete could not work at the beginning of the pandemic, being middle-class and living in a middle-class area of Northern San Diego, Clairemont, meant that Christina did not experience the same level of precarity as Aura. Instead of falling into depression, she released her pent-up energy through home-improvement and art projects, delivering groceries to vulnerable relatives, attending online religious services, and scrolling through Instagram, prompting Pete to accuse her of buying into Q-Anon-related conspiracy theories.

There were others who seemed surprisingly careless during the pandemic. In another part of Clairemont, Catherine found herself chatting about vigilance while sharing a can of Mango White Claw with a white Anglo-American woman, Judith, under illicit fireworks. It was a knowing, guilty transgression for the eight 20- and

30-somethings at this backyard Independence Day gathering. Nacho, the Honduran American host, was well aware of the risks. As a politically left-leaning biotech specialist, he had repeatedly run COVID tests in the lab. In his everyday life, he had been vigilant, not careless. For example, he had criticized the Black Lives Matter protests because he feared that they might become super-spreader events. For him, Independence Day was an exception to his otherwise strict rules. By contrast, many of the Chicanx and Boricua activists that Catherine had attended the protests with refused to celebrate Independence Day, citing the notion that "None of us are free until all of us are free." Under exceptional conditions, excessive expenditure gives community its meaning.[44] When people chose to protest or let down their guard, the losses they risked powerfully proved what they chose to care about. In that way, Independence Day 2020 became an intense celebration of shared identity and collectivity in the face of vulnerability and death – whether or not differently positioned Latinxs celebrated it.

As these examples show, each person's mental and physical health is rooted in their lived experience of intersubjective social and economic conditions and the "cultural expectations of persons in relation to gender, mental and political status."[45] Likewise, each person's experience of the pandemic, their understanding of its causes, and their adherence to the regulations put in place by federal, state, and local government to stem transmission was based on their socio-economic position and previous experiences with manifestations of the state, as well as their political affiliation and the media that they consume. Therefore, whether San Diegans flouted or strongly abided by rules was often a way of expressing different notions of belonging, inspiring different kinds of care and solidarity.

People in white or mixed-ethnic, middle-class areas of San Diego such as Clairemont, often experienced COVID-19 differently than their counterparts in Latinx neighborhoods such as Barrio Logan. As health workers, Nacho and Christina were at considerable risk of being exposed to the virus through their work, but not as much through their contacts, as very few of their neighbors and friends are essential workers. While they and their network have experienced health challenges like recovering from serious traffic accidents, most of them can afford a healthy diet and good health care, which is not the case for people living in Barrio Logan. This meant that one's individual vulnerability and alertness towards the virus was in fact a function of broader structural factors. While "the virus does not discriminate," so that People of Color are not in themselves more likely to

---

44 Martínez, Excess.
45 Jenkins, Introduction, p. 3.

get infected, ethnicity intersects with their class position (and their gender) in ways that determine their health outcomes and differing needs to be vigilant: a middle-class Latina may need to adopt certain watchful practices that middle-class white or Latino men need to use, but not all of the watchful practices that a working class woman in Barrio Logan needs to adopt. As Mexican American journalist Gustavo Arellano remarked to Catherine, "Vigilance is different in middle class areas compared to the barrio. At the end of the day it's about class... Culture does form in you in a way, but if you're rich..."

The pandemic's impact on struggles against gentrification and eco-colonialism in Barrio Logan

Since the beginning of the pandemic, an increasing number of people became unhoused and came to Barrio Logan to access services. As soon as April 2020, CanceltheRents.org posted on their website that

> The working class is experiencing the most unimaginable and devastating crisis of COVID-19, which no one could realistically have prepared for. As we are all aware, millions have lost their jobs, millions more stand to lose their incomes in weeks and months to come. One of the most critical issues is housing.

They argued that the measures instituted by the California State Governor Gavin Newsom to suspend evictions until 31st March 2020 were not nearly enough, since rents still had to be paid when the ban was lifted, and merely suspending evictions just meant the accumulation of an insurmountable level of rental debt. They called on Governor Newsom to declare a statewide cancellation of rents and mortgages and to decree no shutoff of utilities for the course of the pandemic. Comments on the Barrio Bridge Facebook page, which is dedicated to reconnecting the old redlined areas of Barrio Logan, Logan Heights and Sherman Heights, argued that the moratorium on evictions did not go far enough. This included a comment from the owner of a business in Barrio Logan who highlighted that although tenants were not at immediate risk of eviction, they still had to pay rent. The owner said he thought the economic impact of the pandemic would hit small businesses owned by People of Color disproportionately: "This is going to take us back a generation, a decade, and no one is talking about that." He added that it was difficult to know what kind of lasting impact the economic crisis caused by the pandemic would have on businesses and the customers, and worried that it could result in the end of this version of Barrio Logan. He thus directly linked the pandemic and gentrification.

"Gentrification relies on severe urban divestment, which over time, creates 'gentrifiable' building stock, or dirt-cheap real estate."[46] Zukin argues that when city officials attempt to remake urban space they prioritize the ability to consume as an experience, over people's ability to put down roots.[47] The economic value of land and property is prioritized over all else. Launius and Boyce, for example, show that struggles against gentrification combine with struggles against settler-colonialism.[48] In 2009, the Arizona state government took over the board of the Rio Nuevo redevelopment project and threatened to sue both the City of Tucson and Pima County if they did not hand over land of the S'cuk Son site, where the O'odham have lived for over 4 000 years. Despite the majority of Tucson residents voting to protect the land, the state government argued that ecological and cultural preservation would not generate enough economic activity. Indigenous, Chicanx and other residents successfully fought the proposals by forming their own coalition, Rio Nuestro, which was able to effectively protect some of S'cuk Son site from commercial development. However, the Rio Nuevo board still moved forward with mixed-use development, and property values have skyrocketed as the new development has facilitated white gentrification of the West Tucson neighborhood. Sunjata argues that "gentrification within the American context, functions as a more benign form of ethnic cleansing wherein racialized people are evacuated from urban centers; it may be presented as the result of non-violent market forces despite evidence to the contrary [...] Racialized people may develop class consciousness because of the disruptions created by gentrification."[49]

In Barrio Logan, gentrification has been changing the ethnic composition of the neighborhood. The San Diego Union Tribune reported that according to U.S. census data, the proportion of Barrio Logan residents that identified as Hispanic had dropped from 86% in 2000 to 72% in 2010, while the white population expanded by 10% during the same period.[50] According to baseball fan Nacho, the gentrification of Barrio Logan started with building the large parking lot south of the Padres stadium. There had been even more unhoused people in that area before, who were displaced by the construction work. Gentrification intensified in the early 2000s, converging with the global growth in interest in Chicano culture, marked by the success of the *Narcos: Mexico* series and Disney's *Coco*. During that time, Barrio Logan started to attract more tourists, artists, and new residents, and to visibly change. Not all perceived these changes as negative, but others, like the

---

46 Sunjata, Gentrification as settler-colonialism.
47 Zukin, *Naked city*.
48 Launius/Boyce, More than Metaphor.
49 Sunjata, Gentrification as settler-colonialism.
50 Lopez-Villafaña, Barrio Logan community.

132 —— Chapter 5. "Why us?": Making Environmental and Health Threats Visible

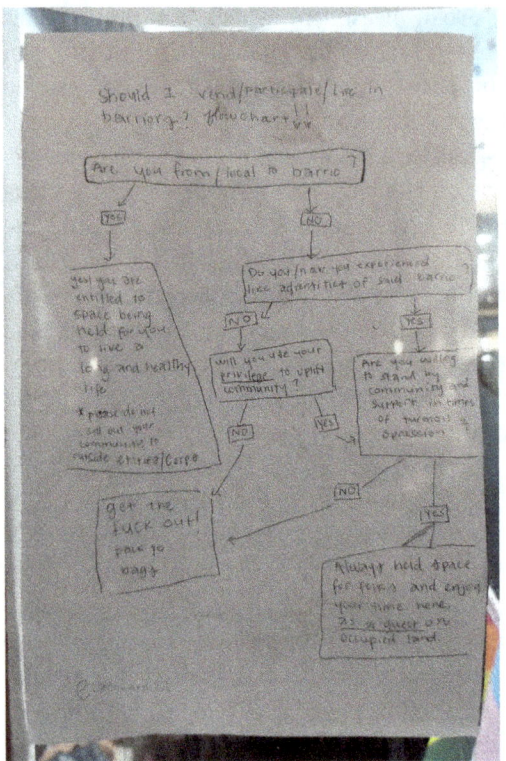

Figure 12: Flow chart on gentrification hanging outside the brujxs' shop.

Witches (see chapter 6) had long been warning against gentrification, as they had been active in helping LGBT Raza affected by it to find safe shelter. A flow chart hanging outside the brujxs' shop (which they gave Catherine permission to photograph) shows their displeasure with the intrusion of outsiders into Barrio Logan and the effect of this on community organizing (Fig. 12).

The Union Tribune article quotes the owner of a local brewery as making a distinction between gentrification and "gentefication," a portmanteau of gentrification and *gente*, the Spanish word for people, which he posited as a desirable alternative: He said the biggest difference between gentrification and "gentefication" is that the latter enhances the characteristics of the neighborhood while at the same time provides economic opportunities for the community. Key to achieving this, he suggested, is "encouraging people in the community to invest in or purchase property."⁵¹ Yet Pepe, a Chicanx student from San Ysidro from a working-class family who has experienced joblessness and becoming temporarily unhoused, saw the notion of "gentefication" critically, pointing out that it benefits middle-class members of the community but displaces the working poor, who cannot afford to pay higher rent, which he viewed as an inevitable consequence of neighborhood "improvements."

Adding further complexity to the situation, two of Catherine's Chicana interlocutors admitted to having contributed to gentrification in their past careers in corporate housing administrations, which both described as soul-crushing work. One of them was Sara, a Mexican and Costa Rican American EHC employee, whom Catherine met for lunch in August 2020. She was wearing angular glasses and a lace mask. After sharing a bit of her life story over tacos, Sara said that a key challenge for the EHC is that some environmental groups led by white people wanted to connect with them but then "flake out of the racial justice work." This dynamic has also been observed in a case where Chicanx and Raza environmental justice activists blocked the construction of a "green energy" power plant in their working-class Los Angeles neighborhoods, finding themselves fighting against former labor rights allies who were promoting a racially neutral pro-jobs campaign. "They were able to block the power plant because their message, oriented to South Gate Latina/o voters, was clear and compelling: ethnic discrimination is bad for our health."⁵² In a different conversation, Berenice, who began working for the San Diego Green New Deal (SDGND) Alliance in 2021, agreed with Sara's view, as she had often seen Indigenous people's visions of land stewardship being marginalized in environmental protection work, a phenomenon that Maya Ch'orti' and Zapotec environmental scientist Jessica Hernandez refers to as "eco-colonialism."⁵³ Native Americans' leadership on environmental struggles has been marginalized for centuries, as they have had to fight the U.S. government on matters of food and water security, and the protection of sacred sites.⁵⁴ One of the latest objects of Indigenous environmental justice struggles has involved fighting the Trump administration's building of a reinforced, higher border wall through sacred burial sites, which has also been destroying century-old saguaro cacti, disrupting native wolves' and jaguars' migratory paths,⁵⁵ beyond being the cause of many migrant deaths.⁵⁶

---

51 Lopez-Villafaña, Barrio Logan community.
52 Brodkin, *Power Politics*, p. 188.
53 Hernandez, *Fresh Banana Leaves*.
54 Gilio-Whitaker, *As long as grass grows*.
55 Ortiz, Trump's border wall endangered ecosystems and sacred sites.
56 De León, *The Land of Open Graves*.

While Sara asserted that environmental racism kills, she nonetheless chose to live in Barrio Logan because she considers it a place of Chicanx resistance and creativity. At the same time, she expressed concern about the ways in which gentrification was changing the neighborhood's identity, while walking along Logan Avenue. Sara said that the barrio used to be a thriving community with two movie theatres and many shops before mixed zoning brought industry there. She said that a well-known Chicano artist used to live in the house where the Border X Brewery is now and they had industry "right next door, blasting toxic fumes into his bedroom." She lamented that the art pipeline for Chicanx artists was gone. There had been a place called Mexihca, where they would invite artists of all levels to contribute to a theme, and even kids participated. If people liked your art there, you would get invited to do a show in the other galleries. Sara was not too sad that another shop had recently closed because it had previously replaced one of those galleries. She then pointed to the herb garden as an EHC achievement. At the time, it was being cared for by the community, but she suggested that EHC might take over again in the future. Finally, they approached a new Italian restaurant which most neighbors like because it is "community-serving," she asserted, although some radicals complain that it is not a Chicanx business. Overall, she concluded that people have mixed feelings about gentrification, so that many vigilantly observe signs of it. They want their families to make a good living but if Chicanxs no longer can afford housing, that will change the barrio. Resonating with her assessment, when Jonathan, Eveline and Carolin took a touristic tour through San Diego in 2021, the guide asserted while quickly passing through Barrio Logan that "now it would be a good time to buy a house" there.

Occasionally, white people join the struggle against gentrification but are often welcomed with skepticism. Many in the community generally "don't trust white people" and are watchful in interactions with them because they can be exploiters, cultural appropriators, infiltrators, or troublemakers, Sara explained. Trust depends on whether someone is willing to vouch for the white person. "So they'll ask, who knows them?" She gave the example of an elder who vouched for a white guy who came to Barrio Logan, wanting to mobilize against gentrification. He collaborated with a new Chicanx business, but ignored the seniority of organizations like Unión del Barrio, who have been working against gentrification and other issues for a much longer time. Sara had a bad feeling about the guy, but the elder insisted that he was just young and well-intentioned. Then, according to Sara, he allowed this guy to sit next to him at the Day of the Dead, in the inner circle where only elders should sit, and later admitted that he had been wrong to take him under his wing, realizing that the guy had kept stirring up trouble and fights. This example shows that even when white people seem conscious about gentrification and want to be allies to Chicanxs, this is much more easily said than done, explaining why many Chicanxs are watchful even in interactions with seemingly well-intentioned white people.

We will now return to the point we began this chapter with: how the extraordinary conditions of the pandemic intensified pre-existing health precarity in the area.

The colonialist militarized apartheid logic of environmental health

Already before the pandemic, there had been multiple reasons why the environment that they live in placed the racialized and migrantized people in Barrio Logan and adjoining neighborhoods at risk of developing underlying health problems. As Jenkins has argued "there is no such thing as individual pathology insofar as the sociocultural milieu profoundly affects the formation of subjective experience and that the core of extraordinary experience is better indexed by struggle than by symptoms."[57] Thus, in Barrio Logan, people's daily struggles against discrimination, crime, over-policing, are connected to ongoing attempts by local people to improve the quality of the air that they breathe. Vigilance against racialized discrimination is mentally taxing, depleting a person's limited psychological resources, leaving them mentally and physically drained.[58]

Maps cited by the EHC, similar to map 3 and map 4 in this chapter, show that pollution levels in San Diego align with the old, redlined districts, including Barrio Logan, showing that not only the land but also the air has a history of segregation. As described in chapter three, in the 1930s greater Logan Heights (as it was at the time) was categorized as a mixed-used zone, meaning industry could be located amongst residential areas. Schools in the community, such as Burbank and Logan Elementary are not even designated as school zones, which would require lower speeds and restrict the passage of heavier trucks.[59] Some companies specifically choose to locate themselves in Barrio Logan because its status makes it a convenient location.

In the recent Youtube video "Que Viva el Barrio: One neighborhood's decades-long fight for a less-polluted future," which discusses the struggles of local people to overturn the categorization of the neighborhood as a mixed-use zone, historian Augie Bareño describes how initially people did not know how best to fight back

---

57 Jenkins, *Extraordinary Conditions*, p. 72.
58 Inzlicht/Schmader, *Stereotype Threat*, p. 166; Crocker, Bodily Imprints of Suffering.
59 Akbany, Environmental Racism and Asthma.

against decisions being imposed on them without consultation which would affect their health on an ongoing basis.[60] However, we have shown that the neighborhood came together as a community following the construction of the Coronado Bridge to create a community space in the form of Chicano Park. Ongoing local struggles have included trying to improve the air-quality of the neighborhood. We can see on map 3 that Barrio Logan as a neighborhood has one of the highest levels of toxic emissions releases in San Diego and map 4 of disadvantaged communities in San Diego shows that across the city, air pollution correlates with deprivation. One strategy involves attempting to change the status of the neighborhood from a mixed-use zone to a residential one. In 2013, the EHC attempted to get a plan passed by the city council that would separate industry from community. It was passed by the city council, but disagreement over where industry stops and the community begins led to the plan being rejected in a city-wide referendum. A new Barrio Logan Community Plan was voted on by San Diego Council on December 7, 2021 and was passed by the council unanimously and sent to the California Coastal Commission for Certification. The plan puts zoning and restricting control in the hands of the community. It aims to create a 9-block buffer zone, separating navy shipyards from homes, thus protecting residents from harmful pollutants released by machinery there.[61]

San Diego's 2019 Climate Action Plan showed that the city council does not allocate resources equitably throughout the city. Not only do neighborhoods like Logan Heights and Barrio Logan have amongst the worst air quality in the city, but are "stuck with the worst commuting prospects, fewest grocery stores, and dirtiest air" and the sidewalks go unfixed.[62] The problem for residents of Barrio Logan, an EHC member wrote in an Opinion article in the San Diego Union Tribune, is that

> The pollution is impossible to escape because it is quite literally everywhere [...] You feel it in your lungs while trying to exercise, or even going for a leisurely walk. [...] The worst part of all this is that our neighborhood did not ask to have all these freeways, industries, and the Port, to be our neighbors. In fact, we fought against it, but the decision was made despite our pushback, and it is we who have had to suffer all these years.[63]

The inconvenient public transportation system means that cars are generally needed to get around the city, which contributes to polluting the air. It is so awkwardly connected that getting from A to B often takes three times longer when taking pub-

---

60 Bareño, Que Viva el Barrio.
61 Akbany, Environmental Racism and Asthma.
62 Pell, Where pollution is worst in San Diego.
63 Garcia, Opinion: In Barrio Logan and Logan Heights, pollution is literally everywhere.

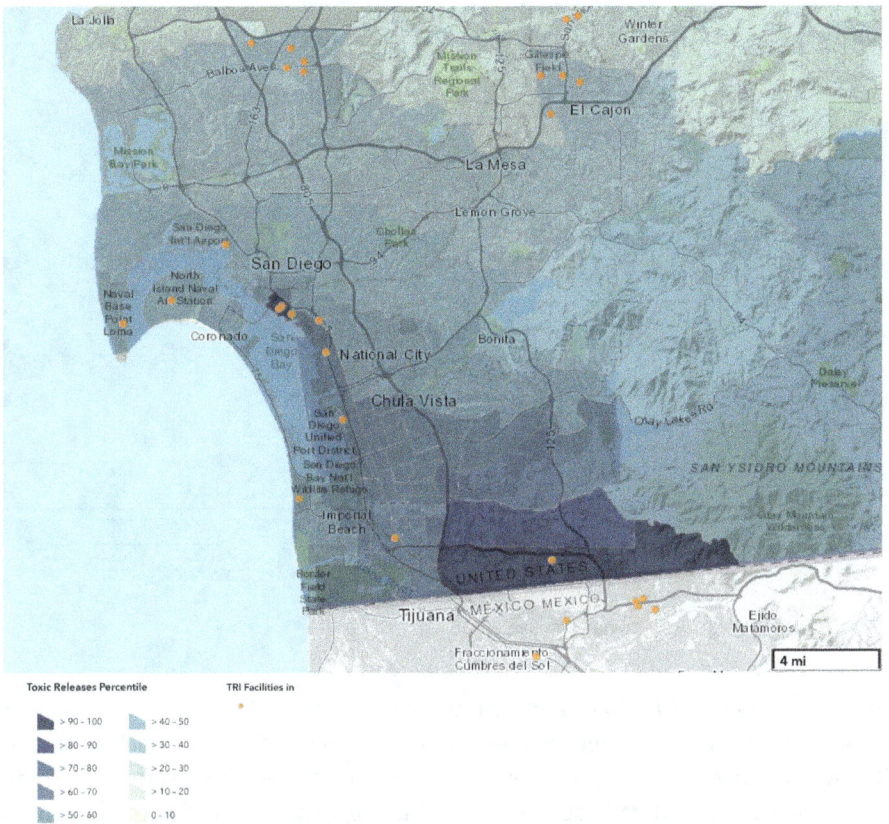

Map 3: Toxic emissions in San Diego according to neighborhood.

lic transport than when driving to the same destination. At the same time, people use their cars even for short distances because they are also a matter of immense pride, and of safety. If you walk by the side of the road then you risk being hit by a careless driver, as one Latina interlocutor lamented who had experienced several near misses on her way home after dark and who was working hard to save enough money to buy a car. Walkers also risk confrontation with pedestrians who cannot afford cars, as Catherine had experienced. Particularly at night, cars are imagined to offer protection from potential criminals. In the Barrio Logan area, research participants often accompanied Catherine to her rental car after dark and sometimes also asked for a text message when she arrived at home. During the pandemic, driving to work additionally shielded commuters from the virus. Yet there was also a strong class component. As a Puerto Rican interlocutor explained, "Cars are so important to Americans because they know they might

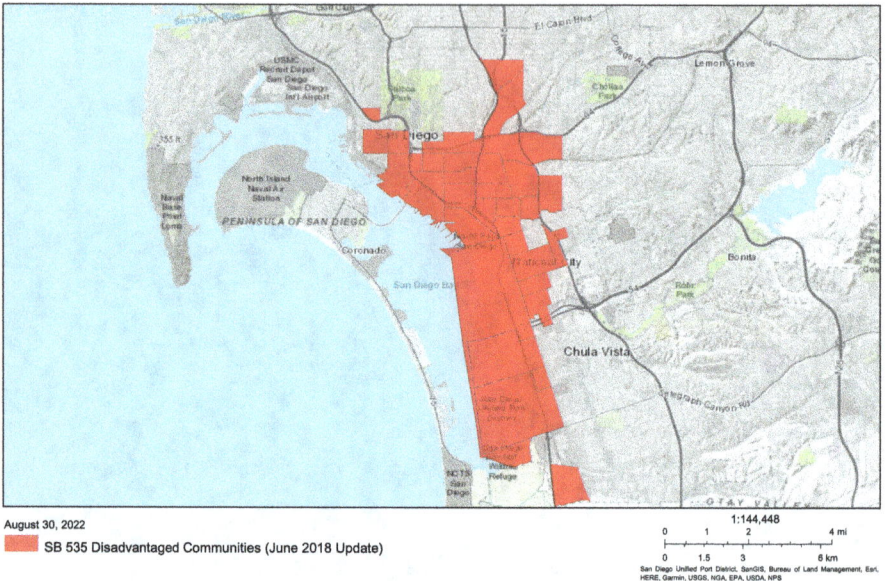

Map 4: Disadvantaged communities in San Diego.

have to live in them. It's like an insurance." A conservative Mexican American man forbade his younger brother from taking public transportation to work, as a matter of pride, given that the public transport system is largely used by the young and the poor, which are often People of Color, which suggests a racialized component to transport self-segregation. Cars are status symbols and many San Diegans strongly identify with their wheels. This is particularly evident in lowrider culture among Chicanxs.[64] Throughout San Diego, if you want to know whether your neighbor is at home, you typically look out of the window to see if their car is there. The people of Barrio Logan are therefore not simply victims, but to some, however small, extent contribute to the pollution problem. However, they have limited choice in the matter, as not having a car in this context means losing the ability to work in most jobs, and therefore also losing face. To prevent their car from being stolen, Chicanx people living in the most crime-afflicted neighborhood of San Ysidro often employ additional security measures such as anti-theft bars for their steering wheel.

This applies also to people like Diego who have worked in the Navy shipyards, thus working for the military-industrial-complex. Shipyard workers implicate themselves not only in racialized pollution but also in the reproduction of the

---

64 Pulido/Reyes, *San Diego Lowriders*.

U.S.-Mexico border as an imagined line that divides and hierarchically differentiates between the Global North and the Global South,[65] a pattern echoed in urban racial segregation, such as through redlining, which creates distinctions between neighborhoods on the basis of class and ethnicity. It should be noted that as an ex-convict without a college degree, Diego had limited job opportunities and had to accept whatever employment he could get. It is likely that many of the locals who work in the shipyards have similarly limited options. As one interlocutor from San Ysidro in his mid-thirties described, he found the working conditions in the shipyards so terrible that he quit after a month.

Notably, "attempts to provide relief to residents of the area have been complicated by the fact that the many maritime businesses in the area, including at the Port of San Diego, vigorously dispute the idea that they're to blame for air pollution. They say the culprit is Interstate 5, which adjoins Barrio Logan."[66] However, from 2019, the state government intended to attribute responsibility by using funding from the state's Community Air Protection Program to place air monitors in Barrio Logan, Logan Heights, Sherman Heights, and West National City.[67] In February 2022, a new program was implemented to help residents affected by the pollution of shipyards and heavy industry, called The Portside Air Quality Improvement and Relief (PAIR) Program. Under the program, over two years, 500 free air purifiers and indoor air monitors would be provided to people living in Barrio Logan, Sherman Heights, Logan Heights, and West National City to support their health in an environment with poor air quality.[68] In these different perspectives on contamination, we can recognize "a clash of temporal perspectives between the short-termers who arrive (with their official landscape maps) to extract, despoil, and depart and the long-termers who must live inside the ecological aftermath and must therefore weigh wealth differently in time's scales."[69]

Catherine's host Pete mentioned to her that he had read a ProPublica piece titled "Welcome to 'Cancer Alley,' Where Toxic Air Is About to Get Worse" about why a certain part of Louisiana has the highest cancer rate in the country. The explanation was the way that predatory companies exploit structural racism, and he found that it was argued with data in a way he found convincing. However, Pete disagreed with Catherine that racism was the root cause for similar problems in Barrio Logan. He seemed much more comfortable acknowledging racism in Loui-

---

65 Besteman, *Militarized global apartheid.*
66 The San Diego Union-Tribune Editorial Board, Will new air pollution study finally lead to relief for Barrio Logan?
67 Ibid.
68 Rendon-Alvarez, Free Air Purifier.
69 Nixon, *Slow Violence and the Environmentalism of the Poor,* p. 17.

siana than in San Diego. Catherine also mentioned the building of infrastructure through poor, non-white neighborhoods, not just in Logan Heights, but across the country. Again, Pete was skeptical, as he has been involved in building freeways and prisons, viewing those as involving complex decisions that cannot be reduced to racism alone. "At least, that's the case today," he corrected himself. He acknowledged that he could not speak for the urban renewal period in the 1950s, a time before his birth.

Sara Ahmed draws on Frantz Fanon to argue that our phenomenological engagement with the world – how we see and experience it – depends on our own circumstances[70]. As Fanon shows, the history of colonialism has made the world 'white', a world in which certain kinds of bodies – white bodies – are deemed to fit. "But non-white bodies do inhabit white spaces; we know this. But such bodies are made invisible at the same time as they become hyper-visible when they do not pass, which means they 'stand out' and 'stand apart.'"[71] While Pete is aware of racism around him, he insists that the law benefits everyone equally. This denial and dismissal of the racist foundations of structural violence is what People of Color must be vigilant of. White fragility, as Fanon describes it, depends on white people's inability to acknowledge the discrimination on which their privilege stands. "This white world is white ignorance, buttressed by a fearful sense that constricts a capacity to feel, our poisoned blood vessels, as to do so will reveal the hypocrisy – the violence, the torture, the genocide – that feeds white lives."[72]

Beyond arguably stemming from a white colonialist[73] and "militarized apartheid" logic,[74] the inequitable distribution of air quality in San Diego demonstrates what Davies calls "the slow violence of pollution."[75] In contrast to Foucault's biopolitics, in which the sovereign power uses the bureaucratic apparatuses of the state to make someone die, necropolitics is used not to actively kill people, but to let them die. People are not being actively killed through pollution as a form of biopolitical control in neighborhoods like Barrio Logan, but "are allowed to suffer the attritional violence of environmental pollution, often through the 'violent inaction' of regulating authorities."[76] Although related to structural violence, Nixon distinguishes slow violence from structural violence since

---

70 Ahmed, Phenomenology of whiteness.
71 Ibid., p. 159.
72 Liebert, *Psycurity*, p. 81, citing Fanon, *Black Skin, White Masks*.
73 Fanon, *Black Skin, White Masks*.
74 Besteman, *Militarized global apartheid*.
75 Davies, Toxic Space and Time.
76 Ibid., p. 1540.

structural violence is a theory that entails rethinking different notions of causation and agency with respect to violent effects. Slow violence, by contrast, might well include forms of structural violence, but has a wider descriptive range in calling attention, not simply to questions of agency, but to broader, more complex descriptive categories of violence enacted slowly over time.[77]

Fundamentally, the slow violence of the pollution affecting the people of Barrio Logan is based on long-standing inequalities that render "some groups more vulnerable to pollution than others."[78] Fanon observed that colonial inequalities have a spatial dimension to them. The colonized and the colonizers live in separate, incommensurate zones and the go-betweens between these different zones are the police and soldiers, who enter the natives' zone with violence.[79] In Barrio Logan, this spatial division is felt most acutely when other San Diegans take decisions that affect their health and quality of life, for example, when the city-wide referendum blocked the Barrio Logan Community Plan in 2013. This shows that Anglo-Americans who disregard that environmental and racial justice are interrelated issues often fail to consult People of Color in decision-making processes, despite People of Color being those most affected by environmental injustice.[80]

As we have described in previous sections, the COVID-19 pandemic intensified some of these issues, but it also alerted people to the interrelatedness of "anti-racism and climate change and indigenous sovereignty and the end of capitalism" and provided activists with the opportunity of envisioning alternatives to it.[81] The example of Berenice's personal and collective struggle for health and environmental justice helps to illustrate these connections.

Watchfulness in extraordinary conditions

Berenice has been active in anti-colonialist, environmentalist, and health organizing, and recognizes the connections between these spheres. While she minimized her infection risk by being highly cautious about masking and disinfecting her hands, Berenice was nonetheless severely affected by the COVID-19 pandemic, as the lockdown disrupted her acupuncture and Chinese medicine doctoral studies. By 2021, at the age of 30, she found herself having to pause her studies

---

77 Nixon, *Slow Violence and the Environmentalism of the Poor*, p. 11.
78 Davies, Toxic Space and Time, pp. 1540 f.
79 Fanon, *The Wretched of the Earth*.
80 Liboiron, *Pollution is Colonialism*.
81 Jiménez Esquinca, *Deabstracting Decolonization*.

indefinitely after running out of student loans but hoped to continue at a later point. For most of the pandemic, she was stuck living in her volatile Anglo American ex-boyfriend's house, who was struggling with alcoholism and jealousy, while she was already burdened with caring for her diabetic mother as well as dealing with her own chronic pain, linked to multiple health conditions as well as potentially being a psychosomatic expression of childhood sexual trauma.

Jenkins argues that what she calls the "extraordinary conditions" of mental illness are fundamentally related to struggles of existential precariousness, which affect People of Color disproportionately.[82] By "extraordinary conditions" she means, firstly, "conditions – illnesses, disorders, syndromes," but secondly also, "conditions – warfare and political violence, domestic violence and abuse or scarcity and neglect of basic human needs – constituted by social situations and forces of adversity."[83] While Jenkins focuses on mental illness, the body similarly interiorizes structural violence, much like a toxin.[84] The particular "extraordinary conditions" of the pandemic experienced by borderlanders like Berenice involves developing a marginalized and precarious subjectivity, which conditions their engagement with the world around them. Anzaldúa refers to borderlanders' sensitive attunement to the world around, resulting in an ever-present fear and awareness of their own oppression, as "la facultad."[85] It is a "kind of survival tactic that people, caught between worlds, unknowingly cultivate."[86] This attunement "is what makes it possible to survive in worlds which are biased against you, or in some cases, actively trying to kill you."[87] *La facultad* also resonates with the constant alertness that Frantz Fanon describes as the colonized adopting to navigate their experience of non-belonging and non-being in a world divided in two.[88]

Having cultivated this intuition for survival meant that Berenice was highly aware of the danger she was in and kept her eyes open for an escape. Things looked particularly bleak for Berenice when she injured herself in 2021 and was thus deprived of her income as a dancer. She seemed trapped with her in many respects well-meaning, but increasingly unhinged and threatening, ex.

This demonstrates the importance of intersectionality for understanding the unequal effects of the pandemic: The COVID-19 virus and lockdowns particularly affected People of Color, and Women of Color even more so. Among working-class

---

82 Jenkins, *Extraordinary Conditions*, p. 249.
83 Jenkins, Introduction, p. 1.
84 Munyikwa, Vigilance as coping.
85 Anzaldúa, *Borderlands/La Frontera*.
86 Ibid., p. 61.
87 Munyikwa, Vigilance as coping.
88 Fanon, *The Wretched of the Earth*.

Latinxs, people of all genders were living in crowded homes, while dealing with the anxiety of high-risk, customer-facing "essential work" and many disruptions to their everyday lives. For *fronterizos* and *fronterizas* like Aura, one of the worst disruptions consisted in no longer being able to cross the border for a long time and being cut off from family members on the other side. Women were additionally burdened with doing unpaid care work at home, and some, like Berenice, were also unable to leave dangerous partners. Worldwide, this forced immobility thus coincided with an uptick in gender-based violence cases.[89]

Therefore, health inequalities, environmental pollution, racialized violence, and gender-based violence are on a continuum that is enmeshed with colonial and military forms of violence. In this way the state and U.S. society's response to the COVID-19 pandemic became a vehicle for enacting racism, territorial domination, and lethal violence. We have shown that the "the slow violence of pollution"[90] in Barrio Logan is predicated on colonialist, militarized conditions in the borderlands[91] and was accelerated by the pandemic, leading to an increase in illness and death among community members during that period. Many survivors of the pandemic were meanwhile driven out of their homes because they had lost work opportunities and could no longer afford their rent, which has also intensified gentrification in the neighborhood.

Fortunately, Berenice's experiences as a community organizer at the Centro Cultural de la Raza and other organizations, and her holistic understanding of health justice landed her a part-time job with the San Diego Green New Deal Alliance. In this role, she was not only working on healing the collective suffering that comes from environmental injustice but was also able to advance her own healing. Getting promoted in her environmental justice job and working full-time finally allowed her to rent a new home for herself, her mother, and her two cats. There, at the beginning of 2022, she planted a native garden and excitedly spoke of hoping to have a baby with her new boyfriend.

Where the government, city planning and even mainstream environmentalists' decisions contribute to racialized and marginalized communities' suffering, many, like Sara and the EHC, react with environmental "counterplanning."[92] As they are aware of living in a settler colonial context, many Chicanxs and other Latinxs like Berenice protest environmental injustice and are working towards a healthier environment. Precisely because watchfulness is their way of life and

---

89 Mittal/Singh, Gender-based violence during COVID-19 pandemic.
90 Davies, Toxic Space and Time.
91 Besteman, *Militarized global apartheid*.
92 Goh, *Form and Flow*.

they alertly perceive the multiple injustices affecting their lives, they insist that "a new way, a new life, a new world is possible."[93]

---

[93] Jiménez Esquinca, *Deabstracting Decolonization.*

Chapter 6
Watchful *Brujxs:* Social Justice Activism in the Digital Sphere

> Firstly, Thank you to [...] fer calling me, [...], in on my internalized anti-blackness and being my hermana through my healing process. Her grace, love, labor, and our conversation helped me open other layers of internalized capitalistic ideologies, abelist ideas, the need to always be strong and perfect etc.
> BRUJXS[1]! YER GUNNA FUCK UUUPP!!! Accept it! This is why shadow work is muy importante!!! To unpack why we get defensive, why are we so afraid to admit we were wrong, accepting that we have so much unconscious biases and internalize anti blackness, racist, colorist, trans antagonistic beliefs, fat antagonistic beliefs, queer antagonistic beliefs, ableist, misogynistic, classiest, and other forms of antagonistic prejudices that is perpetuated throughout society! It's is woven into us throughout transgenerational and intergenerational traumas. We're so unaware of the things we do because we haven't sat down to unpack or listen to the folx that are telling us "what you did was [insert problematic behavior here]" when you do not take time to dismantle your ego, sit with your shadows, and hold space for behaviors your have inherited to survive, you then hurt others because your unwillingness to hold space for yourself to heal. When we learn to be kind and gentle to ourselves, we're able to listen to others when we're told that we're perpetuating problematic ideologies and behaviors. Accountability is ego death. We will be constantly changing the way we speak, move, act, think, and feel in this existence because everyday we will learn something new about our deep programmed existence. The more you're committed to your healing and humbling yourself and move with humility the more the world can heal. Healing work is not about being perfect, it's about decentering your ego so you don't continue the cycle of pain to others and yourself. [...] 🙏🏽
> So where to start, ask yourself "Why am eye so afraid to fuck up?" "Where in my life or in society showed me how a mistake or a perception of mistake has caused significant harm?" "What are the ways eye can unpack to move fearlessly throughout the world so we can dismantle systematic oppression?" (Accept questions to lead to multiple questions, keep unpacking!) The rabbit holes goes fucking deep! Build a support network of friends who are doing the healing work, therapist, folx that hold space for healing, take herbal supports, establish routines that ground and center you, etc. This is a fucking matter of life and death. [...] 🙏🦋
> (06/10/2020, *brujxs*' Instagram page, spelling as in the original text)

This text is from the Instagram page of a group of young self-identified *brujxs* active in San Diego, which they published in June 2020. The opening quote exempli-

---

1 To maintain anonymity for the interlocutors, the group is renamed simply as "the brujxs" here, as was agreed upon with the group itself. In light of the heightened vulnerability of this group, no identifying details are provided in this chapter. *Brujxs*, from *bruja/brujo* (Spanish for witch/witcher) is a non-gendered self-designation that our research partners chose to be used here. While it is commonly used in a derogatory and accusatory way, activist Latinx and Chicanx groups and especially women and non-binary people try to reclaim it.

fies their connection of spiritual and healing work with social justice activism, as well as awareness of oppression (and privilege) of the self and other. The *brujxs* were founded as an activist group and network of three people on a healing journey and as healers, also opening a shop for their own and similar healing products. They offered herbal medicine, self-care products and art, with a particular focus on disadvantaged groups as receivers and producers as well as describing healing modalities and providing workshops. At the same time, the group also opened their Instagram account, through which they have been publishing and (re-)sharing posts and stories that are relevant to them and within their network. These have included support during the healing journey, raising awareness around oppression and marginalization, as well as community support and requests for mutual support. Although the physical shop has since been closed, the group has remained active on markets, in other shops, from their homes and on social media. In the mission statement, which dates to the group's early days and is still available on their Instagram account, the *brujxs* wrote that they are dedicated to creating space, healing and empowerment for their own communities. According to this statement, accessibility for their own community is at the center of their activities and the *brujxs* therefore made their products available on a sliding scale or based on donations. They also decided to donate some of the profits back to their community, as they describe their own position as privileged in comparison to others. The collective presented itself in their "biography" on Instagram as an abolitionist healing queer collective in 2021.

On their Instagram account as well as in interviews with Carolin, the *brujxs* have referred to themselves as healers and witches, sometimes also as medicine healers and energy workers, but they preferred the term *brujxs*. However, the self-designations and identification of the *brujxs* and their network are characterized by fluidity and flexibility, where terminologies and central themes change, adapted to one's own healing process and to the awareness of oppression and marginalization. Thus, the *brujxs* switch between these different terms, while explaining why each of them might not apply in their case. They also identified as queer, women of color, Black and non-binary people of various marginalizations and with Indigenous as well as African roots. The group combined different experiences of migration in the family and although they identified with being Latinx in some ways, they were highly critical of harmful structures in their communities as well. In addition, they also often emphasized the diverse perspectives within their own group due to their different experiences of racism, anti-Blackness and other forms of discrimination and their discrepancies regarding their privilege. The *brujxs*' network on Instagram was at the center of this part of the research, which included people and groups in the research who also deal with healing and consider themselves healers, witches or *brujxs*. These changes in terminology show the difficulties the *brujxs* sometimes had to position themselves and how they tried to avoid categorization while at the same time not being able to avoid it all the time. During her conversations with the *brujxs* and throughout her research, Carolin was expected and felt the need to position herself as well regarding her privileges (see chapter 2). The categories described in this chapter, as well as in the whole book, are not understood as rigid, but changeable and often only expressed in defiance of the same.

Throughout our book, and in particular in this chapter, we stress the heterogeneity and complexity of identity and belonging in the U.S.-Mexico borderlands. Identity and belonging are constantly contested and negotiated within the communities we studied, and self-criticism plays a huge role in most of these resistant groups in the U.S. It is central, for example, to emphasize that Latinxs form a heterogeneous group in terms of their race, ethnicity, and legal status in the U.S. and their experiences are not universal.[2] Nonetheless, many within the community experience systemic oppression that can lead to psychological problems and trauma. As we have emphasized in our study, the Latinx community is often the target of anti-migratory and criminalizing rhetoric, so that many fear and distrust public institutions.[3] It is also important to highlight the experiences, knowledge, and resistance of queer, Indigenous, and Black Latinxs in particular, as they are often rendered invisible in descriptions of the Latinx community[4] and which are at the center of this chapter. Even though they do not fully identify as belonging to the Latinx or Chicanx community, many of the spiritual and activist healing practices that the *brujxs* center are reminiscent of what the Chicanx scholar Irene Lara calls "bruja positionality" by Latinx and Chicanx feminists, who combine the decolonization of their spirituality with their activism,[5] as we will further detail in this chapter.

While the identifiers and subject positions of the *brujxs* may differ from some other protagonists throughout this book, the resistance to coloniality and discrimination in U.S. society plays an important role in their lives, as is the case for Chicanx struggles (see Chapter 3 and 4), and we see this in relation to the coloniality of health (see Chapter 5). Applying an intersectional approach, we are particularly conscious of the need to recognize sameness and difference in relation to power and inequalities.[6] This chapter also provides an example for what decolonization processes, described by Walsh as "the simultaneous and continuous processes of

---

2 Chavez-Dueñas et al., Healing Ethno-Racial Trauma, p. 51.
3 Ibid., p. 54.
4 García, Erased Migrations, p. 3.
5 Lara, Bruja Positionalities.
6 Cho/Crenshaw/McCall, Toward a Field of Intersectionality Studies, p. 795.

transformation and creation, the construction of radically distinct social imaginaries, conditions, and relations of power, knowledge,"[7] could empirically look like.

This chapter's aim is to discuss practices of watchfulness connected with healing practices and the call for social justice. Here, a heightened watchfulness on the one hand preceded practices of healing and awareness for injustice, while on the other hand also being caused by those. The practices towards healing that the *brujxs* offered and used were diverse and included herbal products, meditation, rituals, astrology, crystals etc. Digital watchfulness was most visible in the practices of shadow work and calling out, which both play a part in the post cited at the beginning of this chapter. The *brujxs* centered shadow work as a pillar in the individual healing process, which can happen through self-reflection and self-awareness. Calling out, however, is not understood by the *brujxs* as a constant part of this process, but as a coping mechanism by publicly denouncing those who perpetuate harm within the community and was used sparingly. In this chapter, special attention is placed on the digital sphere as a space of watchfulness and healing. Considering the body of work on "digital vigilantism," which Daniel Trottier understands as "a process where citizens are collectively offended by other citizen activity, and respond through coordinated retaliation on digital media,"[8] this notion can be extended. In the case we present here, digital watchfulness occurs in digital space, considered as overlapping with the physical world as we have described it in Chapters 1 to 5. In line with what we explore ethnographically elsewhere in this book, this includes watchfulness of the self as well as of others in this chapter.

Healing and social justice

Healing was not understood by the *brujxs* as restoring a "normal" or "complete" state that from their point of view cannot be attained. Such an understanding of health is misleading and ableist, in that diseased and neurodiverse bodies are labeled as "not normal," as they explained to Carolin. The healing process is constant and being healed does not exist as a final state, from their perspective.

> And really what is healing? What is it to be healed? I don't even think that is a real thing. [...] I don't know about the term healing, we gotta find a new term because nothing is healed [...].

The members of the group also saw themselves as in the process of healing. They kept the terminologies around healing flexible and even questioned the concept of

---

7 Walsh, "Other" Knowledges, "Other" Critique, p. 11.
8 Trottier, Digital Vigilantism, p. 56.

healing itself, as is evident in the quote above. The *brujxs* understand healing holistically, so it not only includes treating symptoms, but addressing the root of the problem. The roots of the problems were explained as being related to various types of oppression; the acknowledgment of this is considered a further part of the healing process. Healing was aimed at both physical and spiritual well-being by restoring a connection with one's soul, self and ancestors and thus also overcoming the colonization of the self. The *brujxs* explained that this connection was severed by the violent experiences of colonialism, thus marginalizing and oppressing Indigenous healing practices and spirituality:

> And I really think it's just a disconnect from our source and so it's really important that this healing is also holistic as in spiritual not just like physical but that there's like a spiritual connection with our true soul and selves and our ancestors. And you know, our creation stories, you know, because colonization has really removed us from our truths.

The oppression that the *brujxs* and their network had experienced as the root of many (health and mental) symptoms came up frequently during conversations with Carolin. They included racism (and above all anti-Blackness, but also colorism), white supremacy, ableism, patriarchy, queer hostility and especially transphobia, misogyny, classism, (hyper)sexualization, gentrification, othering, (neo)colonialism and capitalism. This list is flexible and is constantly questioned and expanded to include more oppressive systems as needed, as well as acknowledging intersectionality. U.S. society is characterized by systemic and institutionally embedded and maintained inequalities that lead to the marginalization of Black, Indigenous and People of Color, in particular, but also migrant groups and queer and trans people. These and other social and economic inequalities and injustices can be made visible with the concept of intersectionality by looking at intersecting systems of power and their connections to intersecting social inequalities.[9]

Resulting from these systemic oppressions, trauma and general negative mental health effects such as anxiety and exhaustion abound, as we have previously shown in chapter 5. As the U.S. health system, from the *brujxs*' community's point of view, cannot adequately respond to these issues, alternative healing methods together with awareness-building for oppressive structures are seen as central in healing processes. Institutional racism in policy and entrenched in public institutions leads to poorer access to key resources such as education, employment, and health care.[10] In particular, Black (and non-white) women are neglected in the U.S. health care system in this regard.[11]

---

9 Hill Collins, *Intersectionality*, p. 43.
10 Williams/Mohammed, Racism and Health I, p. 1152 f.

Colonialism and coloniality as a root for many of these oppressive systems has been addressed frequently throughout this book. In a similar vein, decolonizing the knowledge and practices of healing was also at the heart of the *brujxs'* activities. Here, the idea of decolonization as an individual as well as collective act refers to reclaiming ancestral knowledge and practices and those of other Black, Indigenous and People of Color from white-dominated spiritual spaces, thereby integrating them into their own lives. The *brujxs* also emphasized the importance of their ancestors in their healing process and practices, as companions and leaders. An example of these acts of decolonial resistance is the self-designation of "witch" and "brujx." Both expressions have negative connotations and are commonly used in a derogatory and accusatory way, however, women, non-binary and trans Black, Indigenous and People of Color in the U.S. reclaim these terms to express their resistance against oppression and to revive Indigenous and African healing knowledge.[12] Modern *brujxs* thus take advantage of the privilege they have by being able to self-identify as such, and according to Norell Martínez, thereby honor those Indigenous and African women who could not openly practice their spirituality and healing practices.[13] While there are also examples where the brujx as a figure is taken more symbolically, for example by being part of popular culture, or as described by Lara[14] as Chicanx and Latinx feminist positionality, the *brujxs* took being a witch literally and used what they summarize as their magic in their community's favor and against the oppressive system. This reclaiming process is not exclusively deemed positive within the Latinx and Chicanx community, where "brujx" and "witch" are commonly still used as a slur and in derogatory and accusatory ways.

This understanding of a connection between spiritual, healing and activist practices relates back to what Latinx and Chicanx scholars have described as well. Lara, for example, argues that some Latinx and Chicanx feminists position themselves as brujx as a form of "spiritual activism." As Lara writes, the experience of dichotomizing and colonizing the sexuality and spirituality of Latinx Women of Color calls for decolonizing them through various healing practices performed by the brujx.[15] Brujx spirituality places connectedness with others and the universe at the center and aims to change the self and others through words, im-

---

11 Bartholomew/Harris/Maglalang, Call to Healing.
12 Martínez, *Bruja Feminism*, p. 231.
13 Ibid., p. 136f. The witch-hunts took place not only in Europe but also in the Americas, where it especially targeted women who were Indigenous healers and those who resisted colonial power (Federici, *Caliban*, p. 231; Silverblatt, *Gender Ideologies*, p. 171).
14 Lara, Bruja Positionalities.
15 Ibid., p. 12.

ages, activism, or healer work.[16] Here Lara takes up the concept of "spiritual activism" from Gloria Anzaldúa and writes that individual as well as communal healing is part of the activity of the brujx:

> The bruja-curandera "spiritual activist" is in the process of embodying and living this spiritual conocimiento, re-membering and creating powerful knowledges for personal and community healing.[17]

In Martínez' view, the historical resistance to the colonial regime in the Americas is connected to the reclaiming of the brujx by female Chicanx, Latinx and Caribbean artists, musicians, writers and others.[18] In doing so, these *brujxs* practice a decolonized spirituality by recalling ancestral ways of knowing, such as Indigenous and African healing knowledge. This reawakens spiritual traditions, understandings of health, the body, nature, and the interconnectedness of all of one's ancestors, and links them to activism for social justice, according to Martínez. Thus, this "bruja spirituality" is also a mode of resistance in the way it is lived out in a non-institutionalized way and in a "do-it-yourself" style. In doing so, however, Martínez also emphasizes that being able to claim the role of witch is a new privilege and honors those African and Indigenous women who could not openly practice their own spirituality or healing practices. The author also mentions that the internet plays a special role for these *brujxs* as a democratizing space and site where they can exercise their resistance and embrace being a brujx. Martínez's work thereby represents a call to Latinx and Chicanx activists in particular to embrace the brujx as a figure who has a feminist, anti-capitalist, and decolonial understanding of life.[19]

More recently, Martínez has argued that these women found new digital presence and growth especially in the times of overt and public misogyny and "white supremacy" under Trump's presidency.[20] Rather than supporting and perpetuating these inequalities, the *brujxs* advocate for social justice in their activities and spirituality. While Martínez also emphasizes that turning to "non-Western" spiritual practices and healing methods is not uncommon or new for racialized people, he explains that the internet and social media in particular gives young individuals

---

16 Pérez, Spirit glyphs, p. 41.
17 Lara, Bruja Positionalities, p. 26.
18 Martínez, *Bruja Feminism*.
19 Ibid., p. xii–239.
20 Martínez, Brujas in the Time of Trump.

a way to share and practice their spiritual practices as well as integrate technology into their spiritual practices.[21]

From the perspective of the *brujxs*, healing and being a healer can only happen in direct connection with the struggle for social justice. For example, in a conversation with Catherine at their shop, they mentioned having participated in pushing back against white supremacist attacks in Chicano Park by casting spells (see chapter 3 for more context on the attacks). In this regard, there are many parallels to the concept of Healing Justice, which expresses the need to firmly incorporate spiritual and healing practices into the activist activities of social movements.[22] While activism against social injustice can itself be partially perceived as healing,[23] queer and trans Black, Indigenous and People of Color in particular describe activist spaces as sites of burnout, rejectionism, and other oppressive structures.[24] For example, Berenice, a young Boricua woman (see chapter 5) avidly engaged in various kinds of healing practices, including herbal Chinese medicine and acupuncture, Western medicine and nutritionist knowledge, restorative justice circles, meditation, yoga, and witchcraft. She felt burnt out both in white-dominated and Chicanx activist spaces, where rigid older male-dominated structures often made young feminist Chicanas and other Women of Color feel unappreciated and ultimately unwelcome in subtle, indirect ways. Due to their own holistic approach to healing, the *brujxs* also engage in social justice activism. The understanding of healing here is linked on the one hand to the anti-ableist normalization of sick, disabled, and neurodiverse bodies, while still providing healing methods if the need is there for them.[25] This understanding also parallels the concept of "extraordinary conditions" we have referred to in Chapter 5, which includes conditions that are defined as mental illness as well as those resulting from adverse social situations in forms of different levels of violence.[26] Janis Jenkins describes the everyday struggles those living in extraordinary conditions experience, hereby also including socially applied stigma.[27] The *brujxs* that are at the center of this chapter similarly interpret healing as a constant process connected to spiritual, emotional, and physical well-being and described it as self-reflexive.

---

21 Ibid., pp. 35–40.
22 Piepzna-Samarasinha, Personal History; Khanmalek, Healing Justice Retrospective.
23 Chavez-Dueñas et al., Healing Ethno-Racial Trauma, p. 60.
24 Piepzna-Samarasinha, Personal History.
25 Ibid.
26 Jenkins, Introduction, p. 1.
27 Ibid., p. 2.

Brujxs and watchfulness on social media

Due to the COVID-19 pandemic, Carolin conducted most of the research digitally. However, focusing on the digital sphere was opportune because the internet and social media have already been described as representing an important site of political participation, networking, interaction, and activism for young people in particular.[28] Social media, and Instagram especially, play an important role in making visible the healing practices of modern self-called witches and are also partially incorporated into their activities.[29] For the *brujxs*, meanwhile, Instagram is more of a necessary evil, as they interact with their collective through the platforms and promote their products and events on it. Moreover, Instagram is also the place where they educate and build awareness within their network and get informed. Their network consists of mostly like-minded people, those who are on a healing journey themselves, potential and past clients, as well as other healers, witches and *brujxs*. According to them, Instagram does not show all of the *brujxs'* activities but is a place of self-expression and activism. The real magic, in the *brujxs* own words, happens in real life. Nevertheless, Instagram plays an important role for the collective: the platform enables the connection to the community and the direct circle of supporters, as well as to people with similar goals. Here, the harm that occurs in physical spaces can slip over into virtual ones. While the second is still a place of coloniality and discrimination, it is one that people can have somewhat more control over. The struggle towards social justice thereby extends to the virtual sphere.

Social media platforms thus become places where solidarity communities are formed, but are also regarded as built on harmful structures that help to maintain the oppression of disadvantaged groups. This happens, for example, in the experience of the *brujxs* and other anti-racist and anti-systemic activists via shadowbanning, which means algorithmically rendering content more invisible[30] or blocking their accounts. Thus, the *brujxs* emphasize that Instagram is not a safe place for disadvantaged and racialized people. Supporting these claims, several studies have found that racism and sexism are inscribed in computer codes.[31] Shadowbanning, but also blocking, difficult accessibility and an often toxic culture of conversation among users make social media an often hostile place for disadvantaged individuals. In the several posts and stories following here, the *brujxs* convey that

---

28 Loader/Vromen/Xenos, The Networked Young Citizen.
29 Martínez, Brujas in the Time of Trump, p. 35.
30 Forsey, Instagram's shadowban.
31 Noble, *Algorithms of Oppression*, p. 9; Raji, Data.

they have been looking for an alternative for a long time but are also still dependent on the platform to enhance their voices and for their enterprise:

> IG [Instagram] wasn't made for us, wasn't made to actually create change. Yes, we can absolutely find one another & share here.
> But the work is happening in our Communities, on the Land, in Ceremony, outside of this app.
> [...] Decolonization & systemic change isn't going to be easy or flowing on an app created to capitalize on us, c3nsor & silence. It happens outside of this space.
> (Post shared by *brujxs*, 05/06/2021).
>
> [...] Look IG does NOT bring me joy. But eye have believed that IG it's the only way to connect with y'all and keep y'all in the loop with *brujxs* to help us stay a float... But eyem starting to feel and see this isn't true... [...]
> (*Brujxs*' post, 04/29/2021)

Many activist groups and community members that experience discrimination use social media as an important place of interaction and reaching out. However, in the same way as discrimination extends into this digital sphere, so does watchfulness. Although digital vigilantism is often motivated by particular understandings of social justice,[32] in practice, the public contests the appropriateness and severity of coordinated retaliation in regard to the offence and their view of participants and targets.[33] This has been referred to as a "parallel form of criminal justice,"[34] involving digital surveillance as well as punishment towards a goal of justice, order, or safety.[35] It is directed towards people and groups who are accused of legal or moral wrongdoing,[36] ranging from "mild breaches of social protocol" to "terrorist acts and participation in riots."[37] However, digital vigilantism has proved difficult to define because aspects such as levels of violence and illegality as well as the relationship to the police, it being an individual or collective, planned or spontaneous act as well as the ultimate goal for it vary greatly.[38] Further, the content of these virtual accusations differs as well, including flagging (shaming a behavior generally), investigating (aims at identifying a person for wrongdoing), hounding (combination of the above involving mobilization against someone) and organized

---

[32] Favarel-Garrigues/Tanner/Trottier, Introducing digital vigilantism, p. 191.
[33] Trottier, Denunciation and doxing, p. 196.
[34] Trottier, Digital Vigilantism, p. 55.
[35] Loveluck, Shades, p. 213.
[36] Trottier, Denunciation and doxing, p. 197.
[37] Trottier, Digital Vigilantism, p. 55.
[38] Loveluck, Shades, p. 214.

leaking (directed at institutions or organizations), as well as forms that are more difficult to categorize.[39]

While these practices can be directed towards social justice and social progression, they can also reproduce misogyny, racism and other discrimination as well as privilege and existing power relations.[40] As an example, Dara Byrne draws a connection between "digilante" tactics by internet users against cybercriminals and anti-Black vigilantism.[41] Here, the actions of a scam-baiting social network especially targeting Nigerian criminal behavior directly parallels anti-Black vigilantism in the U.S. since the mid-1700s, which has been characterized by the careful organization, planning and structure of the violence against Black people and other minorities.[42] Similarly, the targeting of individuals based on categories of suspicion, such as "illegal migrant," have become part of digital vigilantism in Europe and the U.S.[43]

At the same time, digital vigilantism has been described as countering hate-speech and harassment, as well as sexual violence, and has thus been characterized as "progressive" and directed towards social justice.[44] What we argue in this chapter is that using Instagram can become a pathway towards justice and protection for some victim-survivors in the perspective of the *brujxs*, as it is a place where they can assume some control and visibility. We understand this also as a counter-movement against the historical vigilante tactics by racist and nationalist groups in the U.S. that have targeted marginalized people.

In their mission statement, the *brujxs* dedicate themselves to education about racism, anti-Blackness, colorism, anti-trans and -queerness, ableism and misogyny. They used their voice to raise awareness of all types of discrimination in their own communities and also actively fought them. On Instagram, the *brujxs* often expressed their solidarity with other groups worldwide that suffer under oppression and marginalization. They also frequently called for help for individuals in their network who are in financial distress (as mutual aid). However, the reflection of one's own biases, traumatic experiences, and intergenerational trauma common to many disadvantaged communities is also understood by the *brujxs* as an important part of the process of healing. In addition, in Carolin's interviews with the *brujxs*, they often described fear and anger over privileged individuals and corporations whose primary focus is profit co-opting and appropriating their practices.

---

39 Ibid., pp. 217–231.
40 Trottier, Denunciation and doxing, pp. 197, 199.
41 Byrne, 419 Digilantes, p. 72.
42 Ibid., p. 75.
43 Favarel-Garrigues/Tanner/Trottier, Introducing digital vigilantism, p. 189.
44 Ibid., pp. 189–191.

This co-optation and appropriation is understood as an expression of the capitalist system that *brujxs* believe is at the root of many problems in society as a whole, or, in their words: "You shouldn't be able to profit off of the abolitionist activism movement, you shouldn't be able to profit off of activism movements, you shouldn't be able to profit off of healing." While on the one hand acknowledging that their activities are part of "the system," for example referring to capitalism, the *brujxs* fight it by denouncing it and with alternative exchange methods, such as offering their products on sliding scale or donation-based, as well as mutual aid initiatives.

Due to these experiences of and awareness about social injustices, the *brujxs* and their network are watchful against harmful behaviors, structures of oppression and internalized prejudice. Furthermore, due to intergenerational trauma, such as the trauma of enslavement, migration, discrimination and economic hardships, many members of disadvantaged communities do not have trust in the U.S. justice system and their institutions. Thus, they react to not being heard, not being believed, and being treated wrongly when calling for justice by employing their own mechanisms for justice and watchfulness. This is shown towards the self, as exemplified here in shadow work. Shadow work is considered to be about holding oneself accountable to the harm that one has perpetuated as well as acknowledging one's trauma. However, digital watchfulness is also shown as a self-defense-mechanism, where harmful behavior of individuals or small groups is being called out within the community and in society at large. Being (or having to be) highly vigilant, we have argued here, is one of the reasons why many people who have intersectional experiences of oppression experience exhaustion.

Shadow work

> [...] When we learn to be kind and gentle to ourselves, we're able to listen to others when we're told that we're perpetuating problematic ideologies and behaviors. Accountability is ego death. We will be constantly changing the way we speak, move, act, think, and feel in this existence because everyday we will learn something new about our deep programmed existence. The more you're committed to your healing and humbling yourself and move with humility the more the world can heal. Healing work is not about being perfect, it's about decentering your ego so you don't continue the cycle of pain to others and yourself. [...] 💗🙏 [...]
> (06/10/2020, *brujxs'* Instagram page)

"Shadow work" is a concept and healing modality that came up often in texts on social media by the *brujxs*, but also by other self-help groups and networks. This snippet shows that shadow work is about self-improvement as well as introspection, self-reflection and healing of the self. In the post, one of the *brujxs* explained about their own experience of being made aware of internalized ableism, anti-Blackness and capitalistic ideologies, that they needed to address with the help of shadow work. The *brujxs* describe shadow work to Carolin as introspective and detective work, with the aim of accepting one's darker side and to bringing it into balance. However, while it is individual work and a central component of the individual healing process as understood by the *brujxs* and their network, it was also interpreted as being essential to collective and structural changes and social justice. As mentioned in the introduction, the *brujxs* understand their work as abolitionist. As such, in late April 2021, they shared a story on Instagram by someone in their network that said: "abolition requires shadow work." Here, abolition is connected to the power of whiteness that needs to be demolished. In addition, the reflection of everyone's own privileges in connection to whiteness is required to abolish these damaging structures from their perspective.

Shadow work was explained during interviews as being about acknowledging the harm one has done and the accountability process that one has to go through as a result. Thus, this also involves acknowledging a "darker" side of the self and searching for the roots of harmful behavioral patterns individually from the *brujxs*' point of view. They shared that the process of healing inevitably leads to holding oneself accountable. Here, the *brujxs* specified that reflecting upon one's own privilege and internalized prejudices such as racism, colorism, misogyny etc. is necessary work for everyone.

> Because we're all perpetuating harm. Each one of us, we all internalized colorism, anti-Black, fatphobias, transphobias, queerphobia, you know every phobia. Everyone. Even when you are those things. And that's what people don't understand, even when you are the most underrepresented person does not mean that you don't inherit all those issues.

Shadow work was thus considered a necessary step in the healing process and expected from all, especially from the *brujxs* themselves. However, while "sitting with one's shadows" is considered to be hard and exhausting work, it can also help to regain power over oneself and one's own behavior without being led by fear and internalized prejudices against oneself and others. Self-reflection and self-awareness have also been described in other chapters in this book as a distinguishing part of Chicanx and Latinx subjectivation (see Chapter 3 and 4). As mentioned previously, Gloria Anzaldúa's notion of a "new consciousness," similarly describes Chicanxs in the U.S.-Mexico borderlands as being targets and perpetrators of colonial violence at the same time and calls for acknowledging and reflecting on this.[45]

---

45 Anzaldúa, *Borderlands/La Frontera*.

The term shadow work is commonly used within the *brujxs'* network, including in the digital sphere. The term could be used for explaining daily rituals, as well as more generally to summarize a difficult time during the healing process or for specific healing practices. The introductory citation is an example for how shadow work can be talked about on Instagram and where it is made accessible to followers. It makes clear that this practice is understood as a constant learning process, where one has to hold oneself accountable constantly. While it may seem that shadow work is directed solely at the self, the *brujxs* explained it as being "vital in creating change in our society" in a workshop invitation from 2019. The *brujxs* repeatedly emphasized their own privileges in comparison to others, for example, regarding their financial status and housing situation in a neighborhood where people have otherwise experienced dislocation due to gentrification.

Furthermore, shadow work was referred to as the processing of trauma. As previous chapters have shown, many people in the *brujxs'* network and direct surroundings have experienced trauma due to systemic oppression. The *brujxs* educated themselves and others about colonialism as a violent source of disconnection from the spiritual and healing traditions of disadvantaged groups. In order to overcome this violent separation, they deem a decolonization of the self and one's environment necessary. According to them, it was central for the *brujxs* to learn their own history and to decolonize the learned history. In addition, the *brujxs* wrote that radical decolonization is important in order to abolish violent systems and to liberate the Black population. In their view, racism and anti-Blackness exist in close interaction with colonialism:

> We have to decolonize ! These systems are violent, ESPECIALLY TO BLACK FOLX!!!! Black liberation is our priority! If Black people are not put in the forefront of your decolonial work, we don't want it! Everything and we mean EVERYTHING COLONIAL IS LINKED TO RACISM AND ANTI-BLACKNESS !!!! We hope yall are on top of your self care because we have systems to demolish!
> (*Brujxs'* post, 04/13/2021)

While the *brujxs* considered education about and awareness of discrimination as a central step towards healing, they also offered possible coping mechanisms for their clients and network. Shadow work is one example for these coping mechanisms, where they reflect upon traumatic experiences and thus address the roots of health problems to start the healing process. This, according to the *brujxs*, leads to self-empowerment where triggers can be neutralized.

Shadow work is thus an example of watchfulness as a reflection of the self and one's actions and position within one's community and as a part of an oppressive system. In general, experiences of oppression can cause a heightened watchfulness within an individual and a community. This is due to always anticipating negative experiences as a result of personal, communal and ancestral trauma. However, the example of the *brujxs* shows that watchfulness can further be heightened by attempts to cope with trauma and marginalization. While building awareness and centering on reflection about oppressive systems and prejudices, people in the process of healing become more watchful of them. As such, shadow work is a coping mechanism at the same time as it can be a further source of exhaustion.

Calling out

In January and February 2022, videos and texts were created and shared across several important Instagram accounts about a person in the community in Barrio Logan who was accused of "grooming" – building trust with minors to manipulate and abuse them. Some of the Instagram posts and stories shared the experiences of the person who had allegedly been victimized, as well as the experiences of other (mostly anonymous) accusations towards the same person. Other stories directly named the perpetrator and called for accountability, as well as asking for their boycott within the community. Within just a couple of days, several community accounts reacted to this by more generally siding with victims of abuse and sexual violence, as well as calling out this behavior as unacceptable within the community. A video by the accused person also circulated on social media, where they responded to this accusation. However, some within the community did not agree with the pressure produced by this on social media, thus bypassing existing community structures. There seemed to be a generational divide on how this was interpreted: While those more comfortable in using social media within the community prioritized siding with survivors of sexualized violence and discrimination, others emphasized the importance of the existing grassroots structures for decisions on exclusion and accusation within the community.

This is what calling out perpetrators of harm, such as sexualized violence or discrimination, might look like on social media. When the *brujxs* partook in instances of calling out, they usually centered the account of those victimized, their experience and explained how to support them. While those victimized were sometimes kept anonymous, in other instances, their spoken or written narratives about the pain and harm that they have experienced was at the center of this virtual calling out. The online community was asked to share the victim's story, to include them in their spiritual practices to help healing as well as to organize financial support for the victim. The perpetrator was on other accounts asked to be held accountable for their action and sometimes fully named. However, calling out might even look like this (from the introductory example to this chapter):

> Firstly, Thank you to [...] fer calling me, [...], in on my internalized anti-blackness and being my hermana through my healing process. Her grace, love, labor, and our conversation helped me open other layers of internalized capitalistic ideologies, abelist ideas, the need to always be strong and perfect etc.[...]
> (06/10/2020, *brujxs'* Instagram page)

In this post, one member of the *brujxs* recounts their experience of being "called in" for their internalized bias, where one of their friends pointed their ableist and anti-Black bias out to them. Calling in here is a private way of making someone aware of the harm they do as a step before publicly denouncing their harmful acts. Making mistakes is, as is recounted in the example, a logical step during the process of unlearning bias and violence. However, what is more important during this is how a person holds themselves accountable for their action and how they react to being called in (or out). A poster sold in the original store on Logan Avenue exemplified this ethos of calling in by expressing that some individuals build walls for self-protection but are scared of removing them again later because they have grown accustomed to hiding behind them. It describes individual post-trauma psychology but can also be applied to the collective psyche in the face of the U.S.-Mexico border wall.

The structural violence that results from the oppressive system affects not only the individual, but also entire communities and "the body social."[46] For this reason, healing always happens in relation to others for the *brujxs* as well, and individual healing is linked to collective healing from their perspective and in their network. This corresponds to the focus on the connectedness of all in *brujx* spirituality, which means that the community is considered healed only when all its members are healed.[47] This is a similar idea to the abolitionist activist slogan "none of us are free until all of us are free," as heard at a protest at the Otay Mesa Detention Center, where unauthorized migrants are imprisoned (see chapter 5). Starting from the healing process of the individual, the *brujxs*, together with their collective of other healers, place the well-being of the community at the center of their activities, while also directly connecting this to societal change. Being a healer is a survival strategy in a society that, according to them, is built on the oppression of them and their ancestors, and decolonization is an aspect of resistance to this. The spiritual is not considered separate from the political here, as Medina also describes in relation to Chicanx spirituality.[48] Thus, the *brujxs* consider their activities to be part of being activists and of striving towards social justice, as well as a path towards col-

---

[46] Barnes/Sered, Introduction, p. 17.
[47] Martínez, *Bruja Feminism*, p. 238 f.
[48] Medina, Communing with the Dead, p. 206.

lective healing. Here, spirituality, self-improvement as well as self-care are placed into the political aim towards a better future.

As such, the betterment of society and the community was the goal of calling out for the *brujxs*. According to them, calling out is the public denunciation of those people who perpetuate harm to the community or individuals in it and act in a damaging manner, such as when perpetuating discrimination based on gender, race, sexuality or after accounts of sexual violence. Calling out can be used as a defense mechanism to protect those who are most vulnerable to discrimination within the community by enhancing their voices and reports. The *brujxs* did this especially by using their own online presence to publicly denounce someone. However, the *brujxs* emphasized that this calling out often follows a "calling in," meaning that calling out often happens after an attempted discussion or privately holding someone accountable as described above.

Instagram played an important role for this process of calling out, as it is a way to enhance voices that are not necessarily otherwise heard. With a fairly large following, the *brujxs* considered themselves to have a responsibility to share accounts of harm that was perpetuated by individuals or groups within their own community toward especially disadvantaged members. Thus, watchfulness is extended to virtual spaces which becomes crucial to visibility as is proposed by digital vigilantism.[49] As shown in the quote below, the interlocutors for this research emphasized that these instances of calling out could be considered their own take on a justice system that has in the past shown itself to be part of harmful structures of discrimination and injustice:

> [...] for us I feel like again it's about us asking the community to hold this person accountable. It's like hey, this person is perpetuating harm, like whatever it is, you're not safe with this person. You need to like call them out. And that's part of the healing because – but in a way that it's not, again it's not carceral, you know what I'm saying? And like what people don't understand is it's not the same, we're not locking these people [up], we're giving these people the ability to heal themselves by taking away their power, you know what I'm saying?

As already alluded to in the introductory vignette to this section, the process of calling someone out has been criticized by other members of the community and neighbors, who make similar critiques to those previously mentioned by scholars: "Public perception of legitimacy of a denunciation may be based more on ideolog-

---

49 Favarel-Garrigues/Tanner/Trottier, Introducing digital vigilantism, p. 189.

ical context, rather than judicial measures such as proportionality or presumption of innocence."[50]

However, the *brujxs* see it as a tool for holding someone accountable by taking away the perpetrator's power in their environment. As such, this is considered part of the healing process for the person who has been called out, as well as for the community and not least for the victims of abuse and oppression at the same time. The *brujxs* also understand "calling out" as part of the process of decolonization, as it demands that a (often privileged) person takes responsibility for their actions, which might otherwise not be prosecuted by official law enforcement.

Nevertheless, the *brujxs* acknowledge that the practice of calling out can also be harmful for their community and neighborhood if its members attack each other. On the one hand, most harmful behaviors have their roots in a system that should be attacked instead. On the other hand, they consider it important to denounce harmful behavior. They are clear that nobody – including themselves – is perfect and everyone is in a healing and learning process. Nevertheless, the process can have positive effects: "Call out and cancel culture are very, can be very profound when we are doing it to call out truly harmful people, you know," they explained to Carolin during an interview. The aim of this calling out is, on the one hand, to point out the person's harmful behavior and to protect other people from them, and on the other hand to express solidarity with those harmed by these activities. However, denouncing should also be able to lead to the person taking responsibility for the damage they have caused. The *brujxs* said that people who show solidarity with victims and denounce harmful behavior within a community are often perceived as a disruptive factor, but they see it as an opportunity to heal from these problems as a collective.

Ongoing watchfulness and healing for social justice

As the *brujxs* write in their post that we cited at the beginning of this chapter, the healing process of a person and community can only happen through awareness and self-reflection, as well as accountability taken by individuals for the harm they perpetuate. In their perspective:

> We're so unaware of the things we do because we haven't sat down to unpack or listen to the folx that are telling us "what you did was [insert problematic behavior here]" when you do not take time to dismantle your ego, sit with your shadows, and hold space for behaviors your

---

50 Ibid., p. 191.

have inherited to survive, you then hurt others because your unwillingness to hold space for yourself to heal.
(06/10/2020, Instagram post by the *brujxs*)

Digital vigilantism has emerged from a long history of vigilantism, which usually targets more vulnerable groups in society, such as the Black population and unauthorized migrants at the U.S.-Mexico border, as well as other racialized and migrantized people. Similar to this, digital vigilantism often reproduces stereotypes, racism and misogyny. In the case of the *brujxs* discussed here, it is perceived as defiance towards oppression, existing power structures and against those who are perpetrating violence and harm. The digital watchfulness of the *brujxs* and those close to them is directed towards social justice for those who do not feel protected by U.S. society and law enforcement. Calling out is directed outward, where those who have committed harm towards someone within the community are denounced. At the same time, this is about protecting one's one community and using the digital space as a facilitator of visibility. Shadow work, on the other hand, is more directed towards the inside, where one's own part in social injustices and one's privileges are reflected upon. Here, social media is thus a place of awareness building, motivation, self-help as well as a facilitator of interaction with likeminded people. It is also used to tell one's own story and healing journey and as such is a place of subjectivation. While the digital sphere becomes a "safe" space in some ways, where the *brujxs* have their network and "bubble" to interact with, they also acknowledge it being a space that facilitates the reproduction of racism, misogyny and other discriminatory practices.

This chapter emphasizes that practices of digital watchfulness can become a central focus of disadvantaged groups. On the one hand, this results from experiences of oppression and injustice. On the other hand, a focus on calling out and shadow work as exemplified here can lead to higher levels of watchfulness. While the two are different practices, as explained in this chapter, the connections between calling out and shadow work are evident: Both are based on the premise that healing can only happen in direct relation to social justice. In addition, from the perspective of the *brujxs* and other activist groups, social justice can only happen if community members call out harmful behaviors and acts, in order to hold individuals and groups accountable.

Conclusion
Decolonial Watchfulness and Watchful Lives in San Diego

On September 18th, 2019, the former President of the United States, Donald Trump, went down to the Pacific beach near the San Ysidro port of entry to give an update on construction work intended to fortify and improve "the wall" separating the twin cities of San Diego, California, and Tijuana, Baja California Norte. He mentioned that the wall had now been wired, "[…] so that we will know if somebody tries to break through, and you may want to discuss that a little bit, General."

"Sir," the General replied dryly, "there may be some merit in not discussing that."[1]

As this scene illustrates, it is not only surveillance and immigration deterrence that has been intensifying at the U.S.-Mexico border, but also vigilance, like that of the General, without which the effectivity of surveillance technology would be limited. The surveillance and vigilance of the state and of anti-immigration vigilantes in turn provokes more vigilance, more watchfulness, among those who sense that they have been targeted by state surveillance and are being watched.

Racialized and migrantized people living in the U.S.-Mexico borderlands develop practices of watchfulness in response to the increasingly militarized border space they inhabit and the broader colonialist, racial and class dynamics of borderland life. Through our ethnographic research we have argued that the processes involved are very specific to the dynamics of borderland life. This is because, as many scholars of borders have noted, the border is not just a physical line in the sand, but is itself a process,[2] which is "the result of specific relations between power and space."[3] Bordering processes occur in social interactions across various physical spaces in which some bodies are marked as legitimate and others as unwelcome.[4] These processes thus create social boundaries.[5] The border is diffuse,[6] and is present where there are power relationships in which one person (or institution) indicates that another does not belong. The unequal power relationships

---

1 NBC News, Border Wall.
2 Cf. Nail, *Theory*, p. 2; Van Houtum/Van Naerssen, Bordering, ordering and othering.
3 Casaglia, Interpreting the Politics of Borders, p. 28.
4 Johnson et al., Interventions, p. 61.
5 Fassin, Introduction: Connecting Borders and Boundaries, p. 8.
6 Scott, Introduction, pp. 6–8.

Open Access. © 2023 the author(s), published by De Gruyter. This work is licensed under the Creative Commons Attribution 4.0 International License. https://doi.org/10.1515/9783110985573-012

racialized people experience in San Diego are underpinned by the coloniality of the U.S.-Mexico border itself,[7] which made people outsiders in their own land.

Watchfulness is a direct response to this bordering process, but is also itself a bordering process that racialized people employ to articulate themselves as individuals and communities. As borders come to signify cultural boundaries and cultural identity comes to be felt as a core element of one's existence, watchfulness becomes an existential necessity guiding many people's actions, thinking, and presentation of self, particularly in terms of resisting the threat of being assimilated into another way of life. In the process, watchfulness itself becomes a way of life. Therefore, as watchfulness begets watchfulness, it becomes central to borderland subject formation.[8] Yet rather than automatically reproducing itself, the watchfulness of individuals requires activation through direct and indirect calls to place their senses and their attention in service of protecting the collectivity, whether that includes the nation, or a marginalized group. In San Diego, unspoken appeals to public vigilance include building and reinforcing the border wall, or creating political murals. By identifying the centrality of watchfulness to bordering processes, we contribute to debates on this concept in border studies.

On the level of everyday experience, a watchful disposition is informed by the daily experience of racism, coloniality, and the expectation of encountering it, and through the daily insecurities provoked by the feeling of not being protected by the law.[9] Thus, it characterizes Chicanx, Boricua, and other Black, Indigenous, and People of Color's struggles to counteract structures that underpin daily life and the decisions of institutions that exclude or do not fully include them – decisions that entrench coloniality into the way that the city is organized. We argue through our ethnography that knowledge of borderland communities is developed through self-reflexivity about their own discrimination at the hand of the representatives of the state and how these power relations have developed historically. We show power relations being contested through racialized borderland communities trying to change the way that space is used, not as a place of domination and oppression but of creative expression. Most notably we describe the Coronado Bridge as a site of particular power relations which are contested through the creation of Chicano Park in the space below the bridge. We also show the *brujxs* (see chapter 6) to use digital platforms to fight for social justice, self-empowerment and as a tool for vigilance against oppression. With their practices of vigilance in the digital space, they aim to start a healing process towards achieving social justice. Their struggles

---

7 Hernández, *Coloniality of the US/Mexico Border.*
8 Dürr et al., *Becoming Vigilant Subjects.*
9 Goldstein *Outlawed*, p. 122.

against injustice incorporate reflection on one's actions – what the *brujxs* refer to as shadow work. This example thus draws attention to the way in which watchfulness may also contain elements of self-care as a starting point to community care. These "woke" and "careful" elements of the *brujxs*' and other interlocutors' watchful practices radically expand previous conceptualizations of watchfulness.

Yet before we consider the decolonial potential of this concept further, we need to first briefly trace our ethnographic and analytical journey in this book. A key element was the Chicano concept of being *trucha*, which we will describe in the following section, followed by a discussion of Chicanismo as a political movement and identity defined by vigilance – for which we propose the concept of "vigiculture." On this basis, we define watchfulness as a kind of decolonial vigilance and offer some first reflections on how it might be fruitfully employed in support of decolonizing efforts within anthropology.

Trucha!

This book has been informed by the focus of the Collaborative Research Center of "Cultures of Vigilance" at LMU Munich on researching the historical and cultural foundations and variations of vigilance. We have taken on the definition of vigilance used by the CRC as the linking of individual attentiveness, on an everyday basis, to goals set by others. Through the course of the book, we and our racialized interlocutors have used several related words – vigilance, watchfulness, attentiveness – to describe in a nuanced way what many Chicanxs refer to as being *trucha*. This is a novel approach, as watchfulness has not been used to analyze borderland life worlds in Chicano Studies before, and generally represents a theoretically underdeveloped concept in the social sciences.

The significance of the concepts that we use in this book became evident in the conversations with academics and activists in San Diego from the very beginning. In a conversation with Chicano scholar Roberto Hernández in November 2019, he expressed that the park was a great place to study vigilance, but that rather than use the term "vigilance" themselves, Chicanxs of his generation often refer to being *trucha*, a term from Caló (Pachuco/Chicano slang). He lamented that many young people no longer speak Caló. Upon recounting the story of the park, Hernández explained that people are always *trucha* in the park, always keeping an eye on who is coming and going because of various incidents with white nationalists. For this reason, we have focused on the importance of vigilance in Chicano Park. The people of Barrio Logan have displayed aspects of being *trucha* not just in watching out for the dangers around them, but in taking the opportunity – an opportunity they cre-

Figure 13: A tote bag sold on Logan Avenue, Barrio Logan, in August 2022.

ated themselves – to disrupt the oppressive social relationships in which they were caught by creating *their own* social space.

In an email exchange with Alberto Pulido, he confirmed the importance of the concept and wrote that as an academic he would describe being *trucha* as a "coping mechanism" which to some degree is a counter-cultural "attitude of resistance." He also described it as being all about "getting over on the oppressor-strategies of 'movidas' in some cultural contexts." One is *trucha* not just as a fearful act against the violence of one's oppressor, but in watching out for an opportunity to get one over on them.

Henry Kammler has discussed the history of the term as it is used in this context. He points out that hybrid Spanish/English expressions like "staying trucha" or "being trucha" do not have direct equivalents in English.[10] As Kammler describes, the term *trucha* has emerged from variants on non-standard Spanish; the standard Spanish in official use in Mexico and the regional dialects of central Mexico and that of the southwestern U.S. He describes three meanings for *trucha:* firstly, in

---

10 Kammler, Trucha.

standard Spanish a word to describe a species of fish; secondly, in pre-1960s Caló as a word to describe a knife; and finally, to mean "vigilant," "alert," or "smart." This latter meaning Kammler identifies as having emerged in written Mexican sources in the twentieth century, and probably migrated with Spanish speakers to the United States after World War Two.[11]

In this book, we have shown that "being *trucha*" goes beyond adopting a kind of "counter-gaze" in response to what has been referred to as the "white gaze" by Frantz Fanon,[12] or a "double consciousness" by W.E.B. DuBois.[13] As Chicanismo has grown from a different history than African Americans' post-slavery struggles, Chicanxs' experiences of racialization and structural violence partially overlap but also significantly diverge from the realities described by Fanon and DuBois. Being *trucha*, we suggest, forms political subjects, empowers individuals and their communities, as well as aiding their decolonial struggle.

In particular, the creation of Chicano Park in the 1970s as part of their decolonial struggle, alongside the creation of the organizations that look after the park, fed into San Diegan Chicanxs' formation as political subjects. Insofar as the self-representation of Chicanxs through the park's murals has informed subject formation, we follow Jacques Rancière's analysis of the effect of aesthetics on subjectivation in forming a "community of sense."[14] We also add a focus on watchfulness to Fanon's work on colonized subjects' anticipation of the "white gaze" as part of their struggle to free themselves from their subjugation, thus adding a dimension that has not been discussed previously in research on subject formation. As political subjects, Chicanxs face discriminatory structures on a daily basis, and seek to overturn their effects through neighborhood organizations, and this continues to inform their subjectivity. To this extent they could be thought of as "subjects of struggle" (*sujetos de lucha*), a term that the Mexican sociologist Raquel Gutiérrez has used to describe the heterogeneous subjects that come together as protagonists in a collective struggle in producing new forms of cooperation.[15]

The campaign to overturn the categorization of Barrio Logan and Logan Heights as a mixed-use zone, an effect of 1940s redlining, demonstrates that the struggles against racist structural inequality is an ongoing one, and connects the

---

11 Kammler, Trucha, believes that *trucha* most likely entered the lexicon in Mexico from the archaic Iberian Spanish word, truchimán, meaning "a shrewd and astute person that goes about things in a rather unscrupulous fashion" and finds it plausible that *trucha* either entered Mexican Spanish through dialects or was retained in it while falling out of use in peninsular Spanish.
12 Fanon, *Black Skin, White Masks*.
13 DuBois, *The Souls of Black Folk*.
14 Rancière, Contemporary Art.
15 Gutiérrez, *Horizonte Comunitario*.

projects of contemporary residents with their predecessors. When Chicanismo was born as a political movement in the 1960s and 1970s,[16] the project to take more control over the space of the neighborhood became emblematic through the creation of Chicano Park.[17] The younger generation of Chicanxs are aware of past struggles over space and draw on them in contemporary struggles, which focus on gentrification and its alternative, "gentefication" (the uplift of the neighborhood by locals and other Chicanxs and Latinxs rather than by white people moving into the neighborhood, which is often presented as a desirable development, but still displaces poor working-class residents). The present generation's continuing resistance against structural violence is both evidence of the continuing coloniality that racialized U.S. citizens such as Chicanxs face and Indigenous peoples' survivance in the face of genocide.[18]

Present-day Chicanx struggles are connected to, but distinct from those of their forebears, and the evolution of this struggle (and its methodology alongside new technology as well as political and educational advances) has been discussed at length elsewhere,[19] so that we will not go into further detail here. Chicanx watchfulness interweaves temporalities, as can be seen both in the way that contemporary Chicanx political struggles build on those of their forbears, and the embrace of history in Chicano Park. To be Chicanx, as the San Diegan Brown Berets assert, is grounded in not forgetting where one comes from. Chicano Park shows Chicanxs their own history through murals of Chicanx and Latin American icons that have challenged U.S. imperialism, illustrations of United Farmworkers campaigns for better workers' rights – historically led by César Chávez and Dolores Huerta – and above all by the focus on Aztlán, the spiritual project that connects modern Chicanxs with their pre-colonial ancestors. As we quote from Gloria Anzaldúa the *mestiza's* "first step is to take an inventory"[20] of what she inherited from her ancestors. For Anzaldúa, a significant aspect of being Chicanx is the awareness that each individual holds within them Mexican and U.S. American ancestry and the tension between them. Aztlán temporally connects contemporary Chicanxs with a history that incorporates both sides of the present-day border. As a spatial project though it is focused on Chicano Park and therefore invokes guardianship and care on the part of those groups such as the Brown Berets that have been protagonists in the development of the Chicanx movement. Chicanxs' vigilant observa-

---

16 Acuña, *Occupied America*.
17 Kühne/Schönwald/Jenal, Bottom-up memorial landscapes.
18 Vizenor, *Manifest Manners*, p. vii.
19 Vargas, *Crucible of Struggle*.
20 Anzaldúa, *Borderlands/La Frontera*, p. 104.

tion in and around Chicano Park is directed not only towards potential threats from outside, but also towards the behavior of fellow group members.

While we argue that the struggle against coloniality in various guises is what has brought together diverse sets of actors behind Chicanx as a political label, the struggle includes people who do not identify as Chicanxs. Moreover, as San Diego and Barrio Logan are heterogeneous spaces, we have come across overlapping struggles within the same space. For example, what the healers identifying as *brujxs* have in common with people identifying as Chicanxs is their experiences of systematic oppression, and attempts to actively fight it. Self-identified *brujxs* identify the structures of coloniality that affect their lives as sources of trauma. Therefore, in their healing practices, they attempt to address not just the physical body of the individual, but the underlying social injustices.[21] The intersectional nature of struggle in Barrio Logan is emphasized because everyone living there faces the same consequences of coloniality manifested through infrastructure constructed without consultation with the local community, felt most notably through widespread health problems experienced by people living in the highly polluted neighborhood, where people often lack access to adequate health care. We have theorized local activists' awareness-raising campaigns with Anzaldúa's concept of "la facultad,"[22] which describes an embodied attunement to the insidious, largely invisible threat of environmental and health injustice as a form of necropolitics,[23] which was exacerbated by the unequal effects and responses to the COVID-19 pandemic.[24]

To summarize, being *trucha* is a comprehensive concept, in which all the senses become attuned to potential threats, visible and invisible, immediate and structural, in this militarized border context. As a kind of watchfulness that is intimately interlinked with a history of decolonial struggles, it also has caring and spiritual dimensions which offer a potential for healing, which sets it apart from previous conceptualizations of vigilance. For example, Henrik Vigh defined vigilance in terms of negative potentialities only, thus omitting the decolonial potential of being *trucha*, which goes beyond detecting potential or future threats.[25] We have shown that being *trucha* allows Chicanxs to actively shape their future as well as their very sense of self and community. Therefore, this type of vigilance is tied to goals beyond the individual, among which the "community" and "la causa"

---

21 Cf. Jenkins, *Extraordinary Conditions*.
22 Anzaldúa, *Borderlands/La Frontera*, 1987, p. 61.
23 Mbembe, Necropolitics.
24 Cf. Sandset, Necropolitics of COVID-19.
25 Vigh, Vigilance.

(the struggle), appear central. In that sense, Chicanismo does not merely cultivate watchfulness in a culturally specific way, but is defined by it.

From vigilance to vigiculture

Accordingly, watchfulness is not only a "way of seeing," but also "a way of being."[26] Through the *trucha* concept used by Chicanxs, we have presented watchfulness as a "way of life," which is particularly significant in individual subject-formation in the context of the U.S.-Mexico borderlands.[27] As our research has identified, this watchfulness is not just a fearful response to potential surveillance, but a creative attitude that looks for opportunities to challenge the inequalities and disadvantages that they face as a result of coloniality. Such watchfulness, incorporating creative responses to coloniality, has shaped Chicanxs as individuals and collectively, as well as the neighborhood around them.

We suggest the concept "vigiculture" to make visible that vigilance is essential to Chicanismo as a borderland culture, as vigilance is constitutive to the borderlands as a whole. There may be other historical or regional contexts in which vigilance is a temporary cultural feature. What we are emphasizing in this book is that Chicanismo was born from a political struggle for which vigilance was, and continues to be, essential. Therefore, Chicanx lives are watchful lives. The militarization of the U.S.-Mexico border wall in response to a public fear of "criminal migrants" crossing the border itself entrenches and reifies that fear, criminalizes immigration, and prompts ordinary U.S. citizens to watch out for signs of criminality (of others) and criminalization (of oneself) in their environment. This in turn leads to the normalization of racialization and watchfulness in everyday interactions among citizens and with law enforcement officers or other representatives of the state. It follows that, when we write of "militarization," "criminalization," "racialization," and so on, these are relational categories and in this book, we have attempted to show how these categories speak to each other. This is why we have foregrounded watchfulness as contributing to subject formation processes rather than writing about borderland identities in an essentializing mode. A vigiculture is a continuously changing, historically shaped assemblage of practices, terminology, ideologies, and material elements that stem from, or promote, vigilance that are passed on from generation to generation and are necessary to the continuation of the group that identifies with it. Key elements of Chicanismo as a vigicul-

---

26 Finn, Seeing Surveillantly.
27 Dürr et al., *Becoming Vigilant Subjects*.

ture include historical murals reminding people in Barrio Logan to continue fighting for "the cause" or family members teaching each other to stay *trucha* in the face of potential danger.

Yet Chicanismo is not just a vigiculture in a negative sense. Chicanx struggles are tied to practices of care, spirituality, and healing. This means that Chicanxs, like many other racialized and migrantized people, are not just watching out for potential threats, but also looking out for each other, looking within themselves, and looking forward in pursuing decolonial visions. We observed this most clearly at the Women's Circle at the Centro Cultural de la Raza in Balboa Park, which drew on a mixture of Indigenous and other spiritual practices to promote collective healing. While frequently ignored in mainstream theories of power, such care work and spiritual practices are often central in Indigenous Mexican women's conceptualizations of power.[28] They embody, what Sotirin calls feminist vigilance, which combines anger about present circumstances with hopeful and productive visions of a more secure future.[29] We thus suggest that this forward-looking aspect of watchfulness is just as political as the attempt to detect visible and invisible threats in one's surroundings.

One way in which our interlocutors express care lies in cultivating a "wokeness" to the issues surrounding them, including issues that may seem abstract, but that are affecting what they consider their community. For instance, when volunteers at the Centro Cultural create and care for a community garden, they may partially do it for the psychological and spiritual benefits that come from feeling connected with nature and each other. Yet they also assert that they are practicing decolonial resistance by collectively creating spaces of sustainability and self-determination that point toward alternative economies. Similarly, tending to the memorials in Chicano Park is not just an expression of individual mourning, but contributes to the Chicanx struggle, as the collective continuously affirms the dead to be *presente!* (present) at memorial events. In these and other examples, care, wokeness, and watchfulness appear intertwined.

In our conception of vigiculture we draw on Christina Sharpe's metaphor of the "wake," which she employs to make multiple dimensions of Black Americans' struggle for existence visible.[30] The wake at once refers to the path behind the ships transporting slaves across the Atlantic, gatherings to watch the dead before the funeral, and waking up to reality. Sharpe explains how Black Americans' lives continue to be marked by the legacies of slavery – their lives still do not matter to

---

28 Whittaker, Felt power.
29 Sotirin, Introduction to Feminist Vigilance.
30 Sharpe, *In the Wake*.

the dominant culture. Nonetheless, the wake is also a hopeful sign of something that remains and survives in the aftermath of death and suffering. While the historical context of Chicanxs and other Latinxs' struggles is different, the wake's latter aspect of highlighting survivance in the face of the ongoing suffering caused by colonization resonates in the borderland context as well. While grieving and holding vigil for the dead, many of our interlocutors remained defiantly hopeful in their struggle. Borderland watchfulness might then be thought of as a wound rearticulating itself as a decolonial way of being and acting in the world.

Our ethnographic study of watchful behavior thus contributes to studies of vigilance, because it goes beyond individual attentiveness, and includes an element of resistance to the common experience of coloniality that people in Barrio Logan face, thus linking individual attentiveness with working towards a common goal. We have shown this with regards to the historical construction of the Coronado Bridge and with the current struggles of local people for greater control over the environmental conditions of Barrio Logan. The campaigns to improve the quality of life in their neighborhood incorporate elements of being *trucha* because they are engaged in the knowledge that ultimately to create legal changes, Chicanxs and other racialized people have to appeal to the very institutions that discriminate against them.

In summary, while we have highlighted the vigilance of racialized subjects throughout the book, the watchful behavior that we describe is such a prominent feature for many racialized subjects in the U.S. that it could be considered a way of life and even amounts to a "vigiculture" among San Diegan Chicanxs. It is precisely this ability to alertly perceive the multiple injustices affecting their lives, that allows Chicanxs and Raza individuals to build towards another world,[31] a vision that some express as Aztlán. Notably, there is nothing romantic or simple about the struggle that many of our interlocutors are engaged in, and that we, too, as foreign researchers are implicated in. As the San Diegan Chicano poet and organizer Alurista declared in *El Plan Espiritual de Aztlán* of 1969, self-liberation involves hard work and sacrifice: "Aztlán belongs to those who plant the seeds, water the fields, and gather the crops and not to the foreign Europeans."[32] We have shown that in this context, watchfulness, too, is a kind of self-work and community service that is rarely commented on, but crucial to realizing this collective vision. We hope that future studies will develop this conceptualization of watchfulness further, as it might have the potential to advance current debates surrounding security, consciousness, and decolonization.

---

31 Jiménez Esquinca, *Deabstracting Decolonization*.
32 Sánchez, Homeland, p. 135.

To date, the concepts we focus on in this book have not been widely used or studied in anthropology. We aim to open up anthropological debates on vigilance through our ethnographic explorations of these concepts in this book, situating watchfulness in the individual lifeworlds of our interlocutors in San Diego. We also draw attention to watchfulness as anthropological method. As anthropologists, we have become aware of the attentiveness of our interlocutors through our own attentiveness in the field, and their observations of us as well as ours of them. Our own self-reflexivity as ethnographers helped us to understand the self-reflexivity of the subjects of this book. In the following, we will explore what it might mean to take this approach further: What might a watchful anthropology look like?

Toward a watchful anthropology

The concept of "decolonization" has often been criticized as having come to represent an empty buzzword that serves anthropologists' careers, rather than the struggles of the more structurally disadvantaged people they write about. This is not to disparage brilliant colleagues who have worked hard to amplify the voices of those otherwise often unheard and ignored. Yet affirming – as we do – that our interlocutors are "knowledge-producers in their own right"[33] and centering their voices in our writing does not in itself guarantee decolonial outcomes, if colonial logics of space, time, identity, and knowledge remain in place. Thanks to our teachers in San Diego, this study has been a constant reminder of these important nuances in the global decolonization conversation.

Might it be that a genuinely decolonial anthropology is possible if it is watchful – which is to say, aware and careful with respect to its historical legacy, its present impact and future consequences for people's lives? In our research, we have taken some steps towards this by departing from the old, and yet still highly prevalent, Malinowskian approach of just studying a single community in the colonialist mode of gathering the kind of information that has historically enabled governments to control Indigenous populations.[34] Instead, we have sought to shed light on the structures and processes that have been shaping Chicanx and other racialized borderland identities, in order to denaturalize the vigilance that permeates their lives.

We have also departed from ethnographic tradition by not working as lone researchers, but as a team – even while, in an unfortunate echo of Malinowski's

---

33 Restrepo/Escobar, 'Other Anthropologies', p. 118.
34 See González, Anthropology; Simpson, *Mohawk Interruptus*.

forced isolation during fieldwork, our principal ethnographer found herself stranded in California and physically isolated from other team members because of the COVID-19 pandemic. The way in which this book came to fruition was not a linear process, but involved lengthy discussions between the four of us in order to navigate our differences in expectations and approaches. This was similar to some of the challenging conversations we have observed among our interlocutors in activist organizations. Often, our discussions were driven by principles of care and sustainability, such as they centered around wanting to do right by our interlocutors, academic collaborators, and the discipline. All in all, we were able to use teamwork to our advantage both during the joint fieldwork and in the writing process. The joint nature of different aspects of this work meant that we had to make compromises and gained new productive insights through that.

Finally, an obvious, but nonetheless significant difference in our approach compared to Malinowskian anthropology consisted in including digital data. While the physical world was at the center of our book, we hope to inspire further research on watchfulness and social justice vigilantism in the digital sphere. Digital anthropology is a growing field of study that can support sustainability efforts, as it does not require travel, and is therefore more accessible to many researchers (who may lack funding, access to travel visas, or a healthy body). Building research relationships digitally allows us to remain in touch with interlocutors and easily share open access publications from our research while sitting at opposite ends of the world. This allows for discussions to bloom, mistakes to be called out, and for all of us involved in this research to continue looking out for each other.

References

Aaronson, Daniel/Hartley, Daniel/Mazumder, Bhashkar: The Effects of the 130s HOLC 'Redlining' Maps. In: *Working Paper. Federal Reserve of Chicago* (2019).
Abril, Danielle: Drones, robots, license plate readers: Police grapple with community concerns as they turn to tech for their jobs. On: *The Washington Post* (03/09/2022), https://www.washingtonpost.com/technology/2022/03/09/police-technologies-future-of-work-drones-ai-robots/ [last access: 03/09/2022].
Acuña, Rodolfo F.: *Occupied America. A History of Chicanos*. San Francisco 1972.
Aguilera, Jasmine: One Year After Mass Shooting, El Paso Residents Grapple With White Supremacy: 'It Was There the Whole Time'. In: *TIME* (08/03/2020), https://time.com/5874088/el-paso-shooting-racism/ [last access: 05/28/2022].
Ahmed, Sara: A Phenomenology of Whiteness. In: *Feminist Theory* 8/2 (2007), pp. 149–168.
Akbany, Saakib: Environmental Racism and Asthma: Looking Past Barrio Logan as A Public Health Case Study. On: *iHPSD: Intersectional Health Project San Diego*, https://ihpsd.github.io/stories/environmental_racism_and_asthma.html [last access: 10/28/2022].
Alaniz, Yolanda/Cornish, Megan: *Viva la raza. A History of Chicano Identity and Resistance*. Seattle 2008.
Alderman, Jonathan/Goodwin, Geoff: Introduction: Infrastructure as Relational and Experimental Process. In: Alderman, J./Goodwin, G. (Eds.): *The Social and Political life of Latin American Infrastructure. Meanings, Values, and Competing Visions of the Future*. London 2022.
Alderman, Jonathan/Whittaker, Catherine: A Bridge That Divides: Hostile Infrastructures, Coloniality and Watchfulness in San Diego, California. In: *Sociologus* 71/2 (2021), pp. 153–174.
Alonso Bejarano, Carolina/López Juárez, Lucia/Mijangos García, Mirian A./Goldstein, Daniel M.: *Decolonizing Ethnography. Undocumented Immigrants and New Directions in Social Science*. Durham 2019.
Amado, María L.: The "New Mestiza," the Old Mestizos. Contrasting Discourses on Mestizaje. In: *Sociological Inquiry* 82/3 (2012), pp. 446–459.
Amit, Vered: Rethinking Anthropological Perspectives on Community: Watchful Indifference and Joint Commitment. In: Jansen, Bettina (Ed.): *Rethinking Community through Transdisciplinary Research*. Dresden 2020, pp. 48–68.
Anaya, Rudolfo et al.: *Aztlán: Essays on the Chicano Homeland. Revised and Expanded Edition*. Albuquerque 2017.
Anguiano, Marco: The Battle of Chicano Park: A Brief History of the Takeover. In: *Chicano Park Steering Committee*, https://chicano-park.com/cpscbattleof.html [last access: 10/31/2022].
Anthias, Floya: Thinking Through the Lens of Translocational Positionality: An Intersectionality Frame for Understanding Identity and Belonging. In: *Translocations: Migration and social change* 4/1 (2008), pp. 5–20.
Anzaldúa, Gloria: *Borderlands/La Frontera: The New Mestiza*. San Francisco 1987.
Anzaldúa, Gloria: *Borderlands/La Frontera: The New Mestiza*. San Francisco ⁴2012.
Anzaldúa, Gloria: (Un)natural bridges, (Un)safe spaces. In: Anzaldúa, G./Keating, A. (Eds.): *This bridge we call home: radical visions for transformation*. New York/London 2002, pp. 1–5.
Aparicio, Frances R.: (Re)Constructing Latinidad: The Challenge of Latina/o Studies. In: Gutiérrez, R.A./Almaguer, T. (Eds.): *The New Latino Studies Reader*. Oakland 2016, pp. 19–53.
Appadurai, Arjan: *Modernity at Large: Cultural Dimensions of Globalization*. Minneapolis 1997.

Appel, Hannah/Anand, Nikhil/Gupta, Akhil: Introduction: Temporality, Politics, and the Promise of Infrastructure. In: Appel, H./Anand, N./Gupta, A. (Eds.): *The Promise of Infrastructure*. Durham/London 2018, pp. 1–38.

Arellano, Gustavo: Raza Isn't Racist. On: *Los Angeles Times* (06/15/2006), https://www.latimes.com/la-oe-arellano15jun15-story.html [last access: 01/10/2021].

Arfsten, Kerrin-Sina: *The Minuteman Civil Defense Corps: Border Vigilantism. Immigration Control and Security on the US-Mexican Border*. Berlin/Münster 2010.

Arfsten, Kerrin-Sina: Auf der Jagd nach illegalen EinwanderInnen: Aspekte des Grenz-Vigilantismus in den USA. In: *Forum Recht* 1 (2012), pp. 24–27.

Arroyo, Jossianna: "Roots" or the Virtualities of Racial Imaginaries in Puerto Rico and the Diaspora. In: *Latino Studies* 8/2 (2010), pp. 195–219.

Artiga, Samantha/Hill, Latoya/Orgera, Kendal/Damico, Anthony: Health Coverage by Race and Ethnicity, 2010–2019. On: *KFF (Kaiser Family Foundation)* (07/16/2021), https://www.kff.org/racial-equity-and-health-policy/issue-brief/health-coverage-by-race-and-ethnicity/ [last access: 01/06/2022].

Attia, Iman/Zakariya Keskinkılıç, Ozan/Okcu, Büsra: *Muslimischsein im Sicherheitsdiskurs. Eine rekonstruktive Studie über den Umgang mit dem Bedrohungsszenario*. Bielefeld 2021.

Aushana, Christina: Inescapable Scripts: Role-Playing Feminist (Re)Visions and Rehearsing Racialized State Violence in Police Training Scenarios. In: *Women & Performance: A Journal of Feminist Theory* 30/3 (2020), pp. 284–306.

Aviña Cerecer, Gustavo: The Dispossessed of Necropolitics on the San Diego-Tijuana Border. In: *Social Sciences* 9/6 (2020), p. 91.

Balme, Christopher: Hypervigilance: Stayin' Alert and the visual tropes of the war on Corona. In: *Vigilanzkulturen* (05/13/2020), https://vigilanz.hypotheses.org/167 [last access: 10/31/2022].

Barassi, Veronica: Ethnography Beyond and Within Digital Structures and the Study of Social Media Activism. In: Hjorth, L./Horst, H./Galloway, A./Bell, G. (Eds.): *The Routledge Companion to Digital Ethnography*. London/New York 2017.

Barenboim, Deanna: The Specter of Surveillance: Navigating "Illegality" and Indigeneity Among Maya Migrants in the San Francisco Bay Area. In: *PoLAR: Political and Legal Anthropology Review* 39/1 (2016), pp. 79–94.

Bareño, Augie: Que Viva el Barrio: Que Viva el Barrio: One Neighborhood's Decades-Long Fight for a Less-Polluted Future. On: *San Diego Union-Tribune via YouTube* (04/09/2022), https://www.youtube.com/watch?v=B7ApgtzjZ3M [last access: 10/28/2022].

Barnes, Linda L./Sered, Susan S.: Introduction. In: Barnes, Linda L./Sered, Susan S. (Eds.): *Religion and Healing in America*. New York 2004, pp. 3–26.

Barrio Bridge Facebook page: The Chicano Park Steering Committee Objects to Sepolio Early Release. On: *Facebook* (11/04/2020), https://www.facebook.com/298152410237632/photos/a.298647426854797/3699360696783436/ [last access: 10/31/2022].

Barth, Fredrik: Introduction. In: Barth, F. (Ed.): *Ethnic Groups and Boundaries: The Social Organization of Culture Difference*. Bergen 1969.

Bartholomew, Melissa Wood/Harris, Abril N./Maglalang, Dale Dagar: A Call to Healing: Black Lives Matter Movement as a Framework for Addressing the Health and Wellness of Black Women. In: *Community Psychology in Global Perspective* 4/2 (2018), pp. 85–100.

Beer, Bettina: Systematische Beobachtung. In: Beer, Bettina (Ed.): *Methoden ethnologischer Feldforschung*. Berlin 2008, pp. 167–189.

Behar, Ruth: *The Vulnerable Observer: Anthropology That Breaks Your Heart*. Boston 1996.

Beliso-De Jesús, Aisha M./Pierre, Jemima: Anthropology of White Supremacy. In: *American Anthropologist* (2019), pp. 1–11.
Benjamin, Walter: *Stadt des Flaneurs*. Berlin 2015.
Besteman, Catherine: *Militarized Global Apartheid*. Durham 2020.
Blackwell, Maylei: *¡Chicana Power! Contested Histories of Feminism in the Chicano Movement*. Austin 2011.
Blanco, Maria: *A Brief History of the Brown Beret National Organization*. San Diego 1975.
Board, The San Diego Union-Tribune Editorial: Will New Air Pollution Study Finally Lead to Relief for Barrio Logan? In: *Chicago Tribune* (01/30/2019), https://www.chicagotribune.com/sd-barrio-logan-20190129-story.html [last access: 31.10.2022].
Boyer, Dominic: Revolutionary Infrastructure. In: Harvey, P./Jensen, C. Bruun/Morita, A. (Eds.): *Infrastructures and Social Complexity: A Companion*. London/New York 2017.
Brendecke, Arndt: Attention and Vigilance as Subjects of Historiography. An Introductory Essay. In: *Storia della Storiografia* 74/2 (2018), pp. 17–27.
Brendecke, Arndt/Molino, Paola: The Cultures of Vigilance: Historicizing the Role of Private Attention in Society. An introduction. In: *Storia della Storiografia* 74/2 (2018), pp. 11–16.
Briggs, Charles L.: *Learning How to Ask: A Sociolinguistic Appraisal of the Role of the Interview in Social Science Research*. Cambridge 2012.
Brint, Steven: Gemeinschaft Revisited: A Critique and Reconstruction of the Community Concept. In: *Sociological Theory* 19/1 (2001), pp. 1–23.
Brodkin, Karen: *Power Politics: Environmental Activism in South Los Angeles*. New Brunswick, 2009.
Brown Berets National Organization: *Stay Brown*. 1 (April 2021). San Diego 2021.
Browne, Simone: *Dark Matters: On the Surveillance Of Blackness*. Durham 2015.
Butler, Judith: *Kritik der ethischen Gewalt. Adorno-Vorlesungen 2002*. Frankfurt am Main 2007.
Byrne, Dara N.: 419 Digilantes and the Frontier of Radical Justice Online. In: *Radical History Review* 117 (2013), pp. 70–82.
Calderón Gerstein, Walter: COVID-19, Ontopolitics, Necropolitics and a New Philosophical and Social Concept in Perú and the World: Idiopolitics. In: *Comuni@cción* 12/1 (2021), pp. 77–90.
Canclini, Néstor García: *Hybrid Cultures: Strategies for Entering and Leaving Modernity*. Minneapolis 1995.
Casaglia, Anna: Interpreting the Politics of Borders. In: Scott, J.W. (Ed.): *A Research Agenda for Border Studies*. Cheltenham 2020, pp. 27–42.
Casey, Edward S.: How to Get from Space to Place in a Fairly Short Stretch of Time: Phenomenological Prolegomena. In: Feld, S./Bassp, K.H. (Eds.): *Senses of Place*. Santa Fe/Seattle 1996, pp. 13–52.
Castañeda, Heide: *Borders of Belonging: Struggle and Solidarity in Mixed-Status Immigrant Families*. Stanford 2019.
Cattelino, Jessica: Anthropologies of the United States. In: *Annual Review of Anthropology* 39 (2010), pp. 275–292.
Cavanaugh, Maureen/Cabrera, Marissa: Examining The Effects Of Climate Change On Barrio Logan. On: *Kpbs.org* (07/14/2016), https://www.kpbs.org/news/midday-edition/2016/07/14/examining-effects-climate-change-barrio-logan [last access: 10/29/2022].
Centro Cultural de la Raza: *The Chicano Revolt from 1969–1971, Facebook-live symposium*, 01/31/2021, https://www.facebook.com/centrocultural/videos/468979520761542 [last access: 01/18/2023].
Centro Cultural de la Raza: *Raza Visions II: Cultivating Creative Spaces of Autonomy & Resistance. (Zine.)* 1 (2021). San Diego 2021.

Chacón, Justin Akers: *The Border Crossed Us: The Case for Opening the US-Mexico Border.* Chicago 2021.
Chakrabarty, Dipesh: *Provincializing Europe. Postcolonial Thought and Historical Difference.* Princeton 2000.
Chan, Melissa: 'I've Never Seen This Level of Fear.' Why Asian Americans Are Joining the Rush to Buy Guns. On: *TIME* (07/21/2021), https://time.com/6080988/asians-buying-guns/ [last access: 10/28/2022].
Chavez, Leo: *The Latino Threat. Constructing Immigrants, Citizens, and the Nation.* Stanford 2013.
Chávez, Alex E./Pérez, Gina M.: *Ethnographic Refusals, Unruly Latinidades.* Albuquerque/Santa Fe 2022.
Chavez-Dueñas, Nayeli Y./Adames, Hector Y./Perez-Chavez, Jessica G./Salas, Silvia P.: Healing Ethno-Racial Trauma in Latinx Immigrant Communities: Cultivating Hope, Resistance, and Action. In: *American Psychologist* 74/1 (2019), p. 49.
Checker, Melissa: *Polluted Promises: Environmental Racism and the Search for Justice in a Southern Town.* New York 2005.
Chicano Park Steering Committee: Chicano Park Mural Map. In: *Chicano Park Steering Committee* (n.d.), https://chicano-park.com/cpmap.html [last access: 10/28/2022].
Chicano Park Steering Committee. Live-Video. 04.25.2020. https://www.facebook.com/watch/live/?ref=watch_permalink&v=234194984464823. Last accessed 18.01.2023.
Cho, Sumi/Crenshaw, Kimberlé W./McCall, Leslie: Toward a Field of Intersectionality Studies: Theory, Applications, and Praxis. In: *Signs: Journal of Women in Culture and Society* 38/4 (2013), pp. 785–810.
Chu, Julie Y.: When Infrastructures Attack: The Workings of Disrepair in China. In: *American Ethnologist* 41/2 (2014), pp. 351–367.
Cockcroft, Eva: The Story of Chicano Park. In: *Aztlán* 15/1 (1984), pp. 79–102.
Cohen, Anthony P.: Personal Nationalism: A Scottish View of Some Rites, Rights, and Wrongs. In: *American Ethnologist* 23/4 (1996), pp. 802–815.
Collins, Harry: *Tacit and Explicit Knowledge.* Chicago 2010.
Comaroff, Jean/Comaroff, John L.: *Theory from the South, or, How Euro-America Is Evolving Toward Africa.* London 2016.
Cooper Alarcón, Daniel: *The Aztec Palimpsest. Mexico in the Modern Imagination.* Tucson 1997.
Correa, Jennifer G.: The Targeting of the East Los Angeles Brown Berets by a Racial Patriarchal Capitalist State. Merging Intersectionality and Social Movement Research. In: *Critical Sociology* 1 (2011), pp. 83–101.
Couldry, Nick/Mejias, Ulises A.: Data Colonialism: Rethinking Big Data's Relation to the Contemporary Subject. In: *Television & New Media* 20/4 (2019), pp. 336–349.
Council, San Diego City: EXECUTIVE ORDER NO. 2020–11. S.D.C. Council. In: *City of San Diego* (11/22/2020), https://www.sandiego.gov/sites/default/files/executive_order_2020-11.pdf [last access: 10/28/2022].
Crane, Johanna T./Pascoe, Kelsey: Becoming Institutionalized: Incarceration as a Chronic Health Condition. In: *Medical Anthropological Quarterly* 35/3 (2020), pp. 307–326.
Crenshaw, Kimberlé: Mapping the Margins: Intersectionality, Identity Politics, and Violence against Women of Color. In: *Stanford Law Review* 43/6 (1991), pp. 1241–1299. DOI: 10.2307/1229039.
Crocker, Rebecca: Bodily Imprints of Suffering: How Mexican Immigrants Link Their Sickness to Emotional Trauma. In: *The Border and Its Bodies: The Embodiment of Risk Along the U.S.-México Line.* Tucson 2019, pp. 208–236.

Croucher, Stephen M./Nguyen, Thao/Rahmani, Diyako: Prejudice Toward Asian Americans in the Covid-19 Pandemic: The Effects of Social Media Use in the United States. In: *Frontiers in Communication* 5/39 (2020), pp. 1–12. DOI: 10.3389/fcomm.2020.00039.
Cuevas, T. Jackie: *Post-Borderlandia: Chicana Literature and Gender Variant Critique.* New Brunswick 2018.
Davies, Thom: Toxic Space and Time: Slow Violence, Necropolitics, and Petrochemical Pollution. In: *Annals of the American Association of Geographers* 108/6 (2018), pp. 1537–1553.
Davis, Mike/Mayhew, Kelly/Miller, Jim: *Under the Perfect Sun. The San Diego Tourists Never See.* New York 2005.
De La Torre, Renée/Gutiérrez Zúñiga, Cristina: Chicano Spirituality in the Construction of an Imagined Nation. Aztlán. In: *Social Compass* 60/2 (2013), pp. 218–235.
De León, Arnoldo/Griswold del Castillo, Richard: *North to Aztlán. A History of Mexican Americans in the United States.* Wheeling ²2006.
De León, Jason: *The Land of Open Graves: Living and Dying on the Migrant Trail.* Berkeley 2015.
Delgado, Emanuel/Swanson, Kate: Gentefication in the Barrio: Displacement and Urban Change in Southern California. In: *Journal of urban affairs* 43/7 (2021), pp. 925–940.
Delgado, Kevin: A Turning Point: The Conception and Realization of Chicano Park. In: *The Journal of San Diego History* 44/1 (1998), pp. 48–61.
Department of Ethnic Studies: *La Tierra Mia: Chicano Park Murals Documentation Project.* Vol. 1. San Diego 2013.
Domingo Garcia, Maria: I'm a Mother of Four. Palantir's Tech Helped Put Me in an ICE Detention Center. On: *Vice* (12/20/2019), https://www.vice.com/en/article/k7eqwe/im-a-mother-of-four-pal antirs-tech-helped-put-me-in-an-ice-detention-center [last access: 03/10/2022].
DuBois, W.E.B.: *The Souls of Black Folk.* Oxford 2009 [1903].
Dürr, Eveline: *Identitäten und Sinnbezüge in der Stadt. Hispanics im Südwesten der USA.* Münster 2005.
Dürr, Eveline: Feldforschung. In: Beer, Bettina/Pauli, Julia (Eds): *Einführung in die Ethnologie.* Berlin 2017, pp. 89–106.
Dürr, Eveline. ¿Héroe Español O Déspota Colonial? La Controvertida Política De La Memoria En El Sudoeste De Los Estados Unidos De América. In: Gunsenheimer, Antje/Cruz, Enrique Normando/Pallán Gayol, Carlos (Eds.): *El Otro Héroe. Estudios Sobre La Producción Social De Memoria Al Margen Del Discurso Oficial En América Latina.* Göttingen 2020, pp. 165–185.
Dürr, Eveline: Beobachter:in. In: Ege, Moritz/Gozzer, Laura/Habit, Daniel/Wietschorke, Jens/Schwab, Christiane (Eds.): *Kulturelle Figuren. Ein empirisch-kulturwissenschaftliches Glossar.* München [forthcoming].
Dürr, Eveline: Gemeinsames Beobachten als dekoloniale Praxis der ethnologischen Wissensgenerierung. In: Lücking, M./Meiser, A./Rohrer, I. (Eds.): *Im Tandem. Wege zu einer postkolonialen Ethnologie – In Tandem. Pathways towards a Postcolonial Anthropology.* Wiesbaden, 2023, pp. 31–49.
Dürr, Eveline/Alderman, Jonathan/Whittaker, Catherine/Brenner, Christiane/Götze, Irene/Michel, Hannah/Rugel, Agnes/Röder, Brendan/Zelenskaia, Alena: *Becoming Vigilant Subjects.* Hannover 2023.
Dürr, Eveline/Sökefeld, M.: Zeit im Feld: Feldforschung als Paradigma und Praxis. In: Lang, C./ Münster, P./Zehmisch, P./Zickgraf, J. (Eds.): *Ästhetik, Atmosphäre, Medialität: Beiträge aus der Ethnologie.* Münster 2017, pp. 229–238.
Dürr, Eveline/Whittaker, Catherine: Introduction. In: Dürr, Eveline/Whittaker, Catherine (Eds.): *A Multidisciplinary Review Essay of Francisco Cantú's Book, The Line Becomes a River: Dispatches from*

the Border, Vintage, London, 2019. Working Paper des SFB 1369 'Vigilanzkulturen' 3 (2020), pp. 4–6.

Dürr, Eveline/Whittaker, Catherine: "Go Back to Your Country!" Wachsamkeit, Wissen und Kolonialität im US-mexikanischen Grenzraum. In: *Ila. Das Lateinamerika-Magazin* 449 (2021), pp. 4–6.

Dürr, Eveline/Whittaker, Catherine: Wachsamkeit als Alltagspraxis. Dekolonisierung von Zeit und Raum im Chicano Park in San Diego, Kalifornien. In: Brendecke, Arndt/Reichlin, Susanne (Eds.): *Zeiten der Wachsamkeit*. Berlin/Boston 2022, pp. 179–210.

Elenes, Alejandra: Border/Transformative Pedagogies at the End of the Millennium: Chicana/o Cultural Studies and Education. In: Aldama, A./Quiñoz, N.H. (Eds.): *Decolonial Voices. Chicana and Chicano Cultural Studies in the 21st Century*. Bloomington 2002, pp. 245–261.

Emerson, Robert M./Fretz, Rachel I./Shaw, Linda L.: *Writing Ethnographic Fieldnotes*. Chicago ²2011.

Emerson, R. Guy: Vigilant subjects. In: *Politics* 39/3 (2019), pp. 284–299.

Escobar, Arturo: *Designs for the Pluriverse. Radical Interdependence, Autonomy, and the Making of Worlds*. Durham/London 2017.

Fabian, Johannes: *Time and the Other. How Anthropology Makes its Object*. New York 1983.

Falser, Michael S.: Chicano Park. Bürgerinitiative, Graffiti-Kunst und Traumaverarbeitung. Geschichte und Bedeutung von "Chicano Park" in Barrio Logan, San Diego (Kalifornien, USA). In: *kunsttexte.de* 4 (2007), pp. 1–15.

Fanon, Frantz: *The Wretched of the Earth*. London 2001 [1963].

Fanon, Frantz: *Black Skin, White Masks*. New York 2008 [1952].

Farmer, Paul: *Pathologies of Power: Health, Human Rights, and the New War on the Poor*. Berkeley 2005.

Fassin, Didier: Introduction: Connecting Borders and Boundaries. In: Fassin, Didier (Ed.): *Deepening Divides: How Territorial Borders and Social Boundaries Delineate Our World*. London 2020.

Favarel-Garrigues, Gilles/Tanner, Samuel/Trottier, Daniel: Introducing Digital Vigilantism. In: *Global Crime* 21/3–4 (2020), pp. 189–195.

Federici, Silvia: *Caliban and the Witch: Women, the Body and Primitive Accumulation*. New York 2004.

Fernandez, Manny: A Phrase for Safety after 9/11 goes Global. In: *The New York Times* (05/11/2010), https://www.nytimes.com/2010/05/11/nyregion/22slogan.html [last access: 10/31/2022].

Finn, Jonathan: Seeing Surveillantly: Surveillance as Social Practice. In: Doyle, Aaron/Lippert, Randy/Lyon, David (Eds.): *Eyes Everywhere: The Global Growth of Camera Surveillance*. London 2012, pp. 6–80.

Flores-González, N.: *Citizens but not Americans. Race and belonging among Latino millennials*. New York 2017.

Fontein, James: Anthropology and/as Travel. In: *Etnofoor* 9/2 (1996), pp. 5–15.

Fontein, Joost: Doing research: Fieldwork practicalities. In: Konopinski, Natalie (Ed.): *Doing Anthropological Research: A Practical Guide*. Abingdon/New York 2014, pp. 70–90.

Foroutan, Naika: Wie lange bleibt man ein Fremder? Über das Dilemma der Annäherung. In: Felixberger, P./Nassehi, A. (Eds.): *Kursbuch 185: Fremd sein!* Hamburg 2016, pp. 51–66.

Forsey, Caroline: Everything you need to know about Instagram's shadowban. In: *Hubspot* (07/15/2021), https://blog.hubspot.com/marketing/instagram-shadowban [last access: 01/26/2022].

Foucault, Michel: *Discipline and Punish. The Birth of the Prison*. New York 1995 [1977].

Foucault, Michel: The Ethics of the Concern of the Self as a Practice of Freedom. In: Foucault, Michel (Ed.): *Ethics. Subjectivity and Truth*. New York 1997, pp. 281–301.

Foucault, Michel: *On the Government of the Living. Lectures at the Collége de France, 1979–1980*. Basingstoke 2014.

Fránquiz, María E./Ortiz, Alba A.: Co-editors' introduction: Who are the transfronterizos and what can we learn from them? In: *Bilingual Research Journal* 40/2 (2017), pp. 111–115.

Frekko, Susan E./Leinaweaver, Jessica B./Marre, Diana: How (not) to talk about adoption. On communicative vigilance in Spain. In: *American Ethnologist* 42/4 (2015), pp. 703–719.

Fung, Brian: Amazon's Ring has provided doorbell footage to police without owners' consent 11 times so far this year. On: *CNN* (07/14/2022), https://www.cnn.com/2022/07/14/tech/amazon-ring-police-footage/index.html [last access: 08/15/2022].

Galaviz, Manuel Guadalupe: *Expressions of Membership and Belonging. Chicana/o Cultural Politics in Barrio Logan.* Austin 2015.

Garcia, Maritza: Opinion: In Barrio Logan and Logan Heights, pollution is literally everywhere. On: *San Diego Union-Tribune* (10/21/2021), https://www.sandiegouniontribune.com/opinion/commentary/story/2021-10-20/pollution-port-san-diego-barrio-logan [last access: 10/28/2022].

Garcia, Matt: *From the Jaws of Victory. The Triumph and Tragedy of Cesar Chavez and the Farm Worker Movement.* Berkeley 2012.

García, Rocío R.: The Politics of Erased Migrations. Expanding a Relational, Intersectional Sociology of Latinx Gender and Migration. In: *Sociology Compass* 12/4 (2018). DOI: https://doi.org/10.1111/soc4.12571.

Gilio-Whitaker, Dina: *As Long as Grass Grows. The Indigenous Fight for Environmental Justice, from colonization to Standing Rock.* Boston 2019.

Ginwright, Shawn A.: Peace out to revolution! Activism among African American youth: An argument for radical healing. In: *YOUNG* 18/1 (2010), pp. 77–96.

Goh, Kian: *Form and Flow: The Spatial Politics of Urban Resilience and Climate Justice.* Boston, MA 2021.

Goldstein, Daniel M.: *Outlawed: Between Security and Rights in a Bolivian City.* Durham 2012.

Gómez-Quiñones, Juan/Vásquez, Irene: *Making Aztlán: Ideology and Culture of the Chicana and Chicano Movement, 1966–1977.* Albuquerque 2014.

González, Roberto J.: Anthropology, Terrorism, and Counterterrorism. In: MacClancy, Jeremy (Ed.): *Exotic No More: Anthropology for the Contemporary World (Second Edition).* Chicago 2019, pp. 151–176.

Gracia, Jorge J.E.: *Hispanic/Latino Identity: A Philosophical Perspective.* London 1999.

Grasseni, C.: Skilled Visions: Toward an Ecology of Visual Inscriptions. In: Banks, Marcus/Ruby, Jay (Eds.): *Made to be seen: Perspectives on the history of visual anthropology.* Chicago/London 2011, pp. 19–44.

Green, Linda: Fear as a Way of Life. In: *Cultural Anthropology* 9/2 (1994), pp. 227–256.

Gregory, Steven: *Black Corona. Race and the Politics of Place in an Urban Community.* Princeton 1998.

Grosfoguel, Ramón/Hernández, Roberto/Velásquez, Ernesto Rosen: *Decolonizing the westernized university. Interventions in philosophy of education from within and without.* Lanham 2016.

Gupta, Akhil/Ferguson, James: Discipline and Practice: "The Field" as Site, Method, and Location in Anthropology. In: Gupta, Akhil/Ferguson, James (Eds.): *Anthropological Locations. Boundaries and Grounds of a Field Science.* Berkeley 1997, pp. 1–46.

Gutiérrez Aguilar, Raquel: *Horizonte Comunitario. Antagonismo y producción de lo común en America Latina.* Cochabamba 2015.

Gutiérrez, David G.: Historic Overview of Latino Immigration and the Demographic Transformation of the United States. In: Gutiérrez, R.A./Almaguer, T. (Eds.): *The New Latino Studies Reader: A Twenty-First-Century Perspective.* Oakland 2016, pp. 108–125.

Gutiérrez, Ramón A.: What's in a Name? The History and Politics of Hispanic and Latino Panethnicities. In: Gutiérrez, R.A./Almaguer, T. (Eds.): *The New Latino Studies Reader. A Twenty-First-Century Perspective.* Oakland 2016, pp. 19–53.

Gutiérrez, Ramón A./Almaguer, Tomás.: Introduction. In: Gutiérrez, R.A./Almaguer, T. (Eds.): *The New Latino Studies Reader. A Twenty-First-Century Perspective.* Oakland 2016, pp. 1–12.

Gutiérrez, Ramón A./Almaguer, Tomás: *The New Latino Studies Reader. A Twenty First Century Perspective.* Oakland 2016.

Hammad, Lamia Khalil: Border Identity Politics. The New Mestiza in Borderlands. In: *Rupkatha Journal on Interdisciplinary Studies in Humanities* 2/3 (2010), pp. 303–308.

Haraway, Donna: Situated Knowledges: The Science Question in Feminism and the Privilege of Partial Perspective. In: *Feminist Studies* 14/3 (1988), pp. 575–599.

Hartmann, Elke/Jancke, Gabriele: Roupens Erinnerungen eines armenischen Revolutionärs (1921/1951) im transepochalen Dialog. Konzepte und Kategorien der Selbstzeugnis-Forschung zwischen Universalität und Partikularität. In: Ulbrich, Claudia (Ed.): *Selbstzeugnis und Person. Transkulturelle Perspektiven.* Köln 2012, pp. 31–71.

Harvey, Penny: The Materiality of State Effects. In: Krohn-Hansen, Christian/Nustad, Knut G. (Eds.): *State Formation: Anthropological Perspectives.* London 2005.

Harvey, Penny/Jensen, Casper Bruun/Morita, Atsuro: Introduction: Infrastructural complications. In: Companion, A./Harvey, P./Jensen, C.B./Morita, A. (Eds.): *Infrastructures and Social Complexity.* London 2017, pp. 1–22.

Harvey, Penny/Knox, Hannah: *Roads. An Anthropology of Infrastructure and Expertise.* Ithaca 2015.

Hee Lee, Michele Ye: Donald Trump's false comments connecting Mexican immigrants and crime. On: *Washington Post* (07/08/2015), https://www.washingtonpost.com/news/fact-checker/wp/2015/07/08/donald-trumps-false-comments-connecting-mexican-immigrants-and-crime/ [last access: 10/31/2022].

Hernandez, Jessica: *Fresh Banana Leaves. Healing Indigenous Landscapes Through Indigenous Science.* New York 2022.

Hernández, Roberto D.: *Coloniality of the US/Mexico Border. Power, Violence, and the Decolonial Imperative.* Tucson 2018.

Herzog, Lawrence A./Sohn, Christophe: The co-mingling of bordering dynamics in the San Diego–Tijuana cross-border metropolis. In: *Territory, Politics, Governance* 7/2 (2019), pp. 177–199.

Hidalgo, Jacqueline M.: *Revelation in Aztlán Scriptures, Utopias, and the Chicano Movement.* New York 2016.

Hill Collins, Patricia: *Intersectionality as Critical Social Theory.* Durham/London 2019.

Hillis, Ken/Paasonen, Susanna/Petit, Michael: Introduction: Networks of Transmission: Intensitiy, Sensation, Value. In: Hillis, Ken/Paasonen, Susanna/Petit, Michael (Eds.): *Networked Affect.* Cambridge 2015, pp. 1–24.

Hine, Christine: *Ethnography for the Internet: Embedded, Embodied and Everyday.* Milton Park 2015.

Howe, Cymene et al.: Paradoxical Infrastructures: Ruins, Retrofit, and Risk. Science. In: *Technology & Human Values* (2015), pp. 1–19.

Hurst, Alexander: The Vigilante President. On: *The New Republic* (11/06/2019), https://newrepublic.com/article/155579/trump-vigilante-president-supporters-violence [last access: 10/31/2022].

Ibarra, María de la Luz: El Campo. Memories of a Citrus Labor Camp. In: Griswold del Castillo, Richard (Ed.): *Chicano San Diego. Cultural Space and Struggle for Justice.* Tucson 2007, pp. 115–128.

Ingold, Tim: Footprints Through the Weather World: Walking, Breathing, Knowing. In: *Journal of the Royal Anthropological Institute* 16 (2010), pp. 121–139.
Inzlicht, Michael/Schmader, Toni: *Stereotype Threat. Theory, Process, and Application.* Oxford 2012.
Ivasiuc, Ana/Dürr, Eveline/Whittaker, Catherine (Eds.): The Power and Productivity of Vigilance Regimes. In: *Social Conflict. Advances in Research. Special Section* 8/1 (2022), pp. 57–171.
Jenkins, Janis H. (Ed.): *Extraordinary Conditions: Culture and Experience in Mental Illness.* Berkeley/Oakland 2015.
Jenkins, Janis H.: Introduction: Culture, Mental Illness, and the Extraordinary. In: Jenkins, Janis H. (Ed.): *Extraordinary Conditions: Culture and Experience in Mental Illness.* Berkeley 2015, pp. 1–20.
Jiménez Esquinca, Eréndira: *Deabstracting Decolonization: Thoughts from an Individual Exploratory Journey.* San Diego 2021.
Johnson, Corey et al.: Interventions on rethinking the 'border' in border studies. In: *Political Geography* 30/2 (2011), pp. 61–69.
Johnston, Les: What is Vigilantism? In: *The British Journal of Criminology* 36/2 (1996), pp. 220–236.
Jorgensen, Danny L.: *Participant Observation: A Methodology for Human Studies.* Newbury Park 1989.
Kammler, Henry: Trucha: What's So 'Trout' About Being Vigilant? In: *Vigilanzkulturen* (04/08/2021), https://vigilanz.hypotheses.org/1381 [last access: 10/31/2022].
Khanmalek, Tala: Outro: A Healing Justice Retrospective. In: *nineteen sixty nine: an ethnic studies journal* 2/1 (2013), pp. 1–5.
Kim, Nadia Y.: *Refusing death. Immigrant Women and the Fight for Environmental Justice in LA.* Stanford 2021.
Kirszbaum, Thomas: Urban Renewal in the USA: A Neoliberal Policy? On: *Métropolitiques* (05/03/2019), https://metropolitics.org/Urban-Renewal-in-the-USA-A-Neoliberal-Policy.html [last access 10/20/2022].
Klinke, Ian: Chronopolitics. A Conceptual Matrix. In: *Progress in Human Geography* 37/5 (2012), pp. 673–690.
Kucher, Karen/Figueroa, Teri: Driver who killed 4 in Chicano Park crash set for early prison release; DA calls the move 'unconscionable'. On: *The San Diego Union-Tribune* (11/04/2020), https://www.sandiegouniontribune.com/news/public-safety/story/2020-11-04/driver-who-killed-4-in-chicano-park-crash-set-for-early-prison-release-da-calls-the-move-unconscionable [last access: 10/31/2022].
Kühne, Olaf/Schönwald, Antje: *Eigenlogiken, Widersprüche und Hybriditäten in und von "America's finest city".* Wiesbaden 2015.
Kühne, Olaf/Schönwald, Antje/Jenal Corinna: Bottom-up memorial landscapes between social protest and top-down tourist destination. The case of Chicano Park in San Diego (California) – an analysis based on Ralf Dahrendorf's conflict theory. In: *Landscape Research* (2022), pp. 1–17.
Kun, Josh/Montezemolo, Fiamma: Introduction. The Factory of Dreams. In: Kun, Josh/Montezemolo, Fiamma (Eds.): *Tijuana Dreaming. Life and Art at the Global Border.* Durham/London 2012, pp. 1–20.
Laine, Jussi P./Casaglia, Anna: Challenging borders: a critical perspective on the relationship between state, territory, citizenship and identity: Introduction. In: *Europe Regional* 24/1,2 (2017), pp. 3–8.
Lara, Irene: Bruja Positionalities: Toward a Chicana/Latina Spiritual Activism. In: *Chicana/Latina Studies* 4/2 (2005), pp. 10–45.
Larkin, Brian: The Politics and Poetics of Infrastructure. In: *Annual Review of Anthropology* 42/1 (2013), pp. 327–343.
Latorre, Guisela: *Walls of Empowerment: Chicana/o Indigenist Murals of California.* Austin 2008.

Launius, Sarah/Boyce, Geoffrey Alan: More than Metaphor: Settler Colonialism, Frontier Logic, and the Continuities of Racialized Dispossession in a Southwest U.S. City. In: *Annals of the American Association of Geographers* 111/1 (2020), pp. 157–174.

Lemanski, Charlotte: Infrastructural citizenship: The everyday citizenships of adapting and/or destroying public infrastructure in Cape Town, South Africa. In: *Transactions of the Institute of British Geographers* 45/3 (2019), pp. 589–605.

Le Texier, Emmanuelle: The Struggle against Gentrification in Barrio Logan. In: Griswold del Castillo, Richard (Ed.): *Chicano San Diego. Cultural Space and Struggle for Justice.* Tucson 2007, pp. 202–221.

Liboiron, Max: *Pollution is Colonialism.* Durham 2021.

Liebert, Rachel Jane: *Psycurity: Colonialism, Paranoia, and the War on Imagination.* London 2018.

Little, Walter E./Rees, Martha W.: Introduction: Participatory Research and Ethics in Mesoamerican Fieldwork. In: *Annals of Anthropological Practice* 44/2 (2020), pp. 145–151.

Lizárraga, José Ramón/Gutiérrez, Kris D.: Centering Nepantla Literacies from the Borderlands. Leveraging "In-Betweenness": Toward Learning in the Everyday. In: *Theory Into Practice* 57/1 (2018), pp. 38–47.

Loader, Brian D./Vromen, Ariadne/Xenos, Michael A.: The Networked Young Citizen: Social Media, Political Participation and Civic Engagement. In: *Information, Communication & Society* 17/2 (2014), pp. 143–150.

Lopez-Villafaña, Andrea: Barrio Logan community members discuss gentrification. On: *San Diego Union Tribune* (09/30/2019), https://www.sandiegouniontribune.com/communities/san-diego/story/2019-09-28/barrio-logan-residents-discuss-gentrification [last access: 11/14/2022].

Lorde, Audre: *Your Silence Will Not Protect You.* London 2017.

Lorenzini, Daniele/Tazzioli, Martina: Confessional Subjects and Conducts of Non-Truth. Foucault, Fanon, and the Making of the Subject. In: *Theory, Culture & Society* 35/1 (2018), pp. 71–90.

Loveluck, Benjamin: The many shades of digital vigilantism. A typology of online self-justice. In: *Global Crime* 21/3–4 (2020), pp. 213–241.

Lukens, Gideon/Sharer, Breanna: Closing Medicaid Coverage Gap Would Help Diverse Group and Narrow Racial Disparities. In: *Center on Budget and Policy Priorities.* (06/14/2021), https://www.cbpp.org/research/health/closing-medicaid-coverage-gap-would-help-diverse-group-and-narrow-racial [last access: 08/01/2022].

Makena, Kelly: Inside Nextdoor's 'Karen Problem'. On: *The Verge* (06/08/2020), https://www.theverge.com/21283993/nextdoor-app-racism-community-moderation-guidance-protests [last access: 03/10/2022].

Mareš, Miroslav/Bjørgo, Tore: Introduction: Vigilantism against Migrants and Minorities – Concepts and Goals of Current Research. In: Mareš, Miroslav/Bjørgo, Tore (Eds.): *Vigilantism against Migrants and Minorities.* London 2019, pp. 1–30.

Martínez, Norell: *Bruja Feminism & Cultural Production: Reclaiming the Witch in the Neoliberal Era.* San Diego 2019.

Martínez, Norell: Brujas in the Time of Trump. Hexing the Ruling Class. In: Navarro, Sharon A./Saldaña, Lilliana Patricia (Eds.): *Latinas and the Politics of Urban Spaces.* New York/London 2021, pp. 31–52.

Martínez, Samuel: Excess: The Struggle for Expenditure on a Caribbean Sugar Plantation. In: *Current Anthropology* 51/5 (2010), pp. 609–628.

Marx, Jesse: San Diego Smart Streetlights Are Off, But They're Still Helping Police. On: *Voice of San Diego* (12/23/2020), https://voiceofsandiego.org/2020/12/23/san-diego-smart-streetlights-are-off-but-theyre-still-helping-police/ [last access: 08/15/2022].
May, Todd: *Contemporary Political Movements and the Thought of Jacques Rancière: Equality in Action.* Edinburgh 2010.
Mbembe, Achille: Necropolitics. In: *Public Culture* 15/1 (2003), pp. 11–40.
McCaughan, Edward J.: We Didn't Cross the Border, the Border Crossed Us: Artists' Images of the US-Mexico Border and Immigration. In: *Latin American and Latinx Visual Culture* 2/1 (2020), pp. 6–31.
McKenna, Stacey A.: "We're Supposed to Be Asleep?" Vigilance, Paranoia, and the Alert Methamphetamine User. In: *Anthropology of Consciousness* 24/2 (2013), pp. 172–190.
Medina, Lara: Communing with the Dead: Spiritual and Cultural Healing in Chicano/a Communities. In: Barnes, Linda L./Sered, Susan S. (Eds.): *Religion and Healing in America.* New York 2004, pp. 205–216.
Meinhof, Marius: Die Kolonialität der Moderne. Koloniale Zeitlichkeit und die Internalisierung der Idee der "Rückständigkeit" in China. In: *Zeitschrift für Soziologie* 50/1 (2021), pp. 26–41.
Mendes, Kaitlynn/Ringrose, Jessica/Keller, Jessalynn: *Digital Feminist Activism. Girls and Women Fight Back Against Rape Culture.* Oxford 2019.
Méndez, Michael: *Climate Justice from the Streets: How Conflict and Collaboration Strengthen the Environmental Justice Movement.* New Haven 2020.
Metzl, Jonathan Michel: *Dying of Whiteness. How the Politics of Racial Resentment is Killing America's Heartland.* New York 2019.
Mignolo, Walter D./Tlostanova, Madina V.: Theorizing from the Borders. Shifting to Geo- and Body-Politics of Knowledge. In: *European Journal of Social Theory* 9/2 (2006), pp. 205–221.
Mignolo, Walter D./Walsh, Catherine E.: *On Decoloniality. Concepts, Analytics, Praxis.* Durham/London 2018.
Miller, Todd: *More Than A Wall: Corporate Profiteering and the Militarization of US Borders.* Amsterdam 2019.
Mills, Charles W.: The Chronopolitics of Racial Time. In: *Time & Society* 29/2 (2020), pp. 297–317.
Mitchell, Timothy: Society, Economy, and the State Effect. In: Sharma, Aradhana (Ed.): *The Anthropology of the State: A Reader.* Malden 2006, pp. 76–97.
Mittal, Shalini/Singh, Tushar: Gender-Based Violence During COVID-19 Pandemic: A Mini-Review. In: *Front. Glob. Womens Health* 1/4 (2020). DOI: 10.3389/fgwh.2020.00004 [last access: 10/31/2022].
Munyikwa, Michelle: Vigilance as coping, vigilance as injury. On: *Somatosphere* (03/06/2019), http://somatosphere.net/2019/vigilance-as-coping-vigilance-as-injury.html/ [last access: 10/31/2022].
Muñoz-Hunt, Toni: Aztlán: From Mythos to Logos in the American Southwest. In: *Borders in Globalization Review* 1/1 (2019), pp. 54–65.
Nail, Thomas: *Theory of the Border.* Oxford 2016.
NBC News: President Donald Trump Tours Border Wall: 'Now We Have A World Class Security System'. On: *NBC News, YouTube.com* (09/19/2019), https://www.youtube.com/watch?v=qJLTEmfSuTU [last access: 09/08/2022].
Nevins, Joseph: *Operation Gatekeeper and Beyond. The War on "Illegals" and the Remaking of the U.S.-Mexico Boundary.* New York/London 2010.
Nielsen, Cynthia R.: *Foucault, Douglass, Fanon, and Scotus in Dialogue. On Social Construction and Freedom.* London 2013.
Nixon, Rob: *Slow Violence and the Environmentalism of the Poor.* Cambridge 2011.

Noble, Safiya Umoja: *Algorithms of Oppression: How Search Engines Reinforce Racism.* New York 2018.
Norris, Frank: Logan Heights: Growth and Change in the Old 'East End'. In: *The Journal of San Diego History* 29/1 (1983), pp. 28–40.
Oliver, Pamela: Race Names. On: *Race, Politics, Justice* (09/16/2017), https://www.ssc.wisc.edu/soc/race politicsjustice/2017/09/16/race-names/ [last access: 10/31/2022].
Oosterbaan, Martijn/Jaffe, Rivke: Popular Art, Crime and Urban Order Beyond the State. In: *Theory, Culture & Society Preprint* (2022), pp. 1–20.
Ortiz, Erik: Trump's border wall endangered ecosystems and sacred sites. Could it come down under Biden? On: *NBC News* (11/11/2020), https://www.nbcnews.com/science/environment/trump-s-border-wall-endangered-ecosystems-sacred-sites-could-it-n1247248 [last access: 08/01/2022].
Ortiz, Isidro D.: "¡Sí, Se Puede!" Chicana/o Activism in San Diego at Century's End. In: Griswold del Castillo, Richard (Ed.): *Chicano San Diego. Cultural Space and Struggle for Justice.* Tucson 2007, pp. 129–157.
Palacios, Agustín: Multicultural Vasconcelos: The Optimistic, and at Times Willful, Misreading of La Raza Cósmica. In: *Latino Studies* 15/4 (2017), pp. 416–438.
Pell, Sheila: Where pollution is worst in San Diego. On: *San Diego Reader* (02/18/2020), https://www.sandiegoreader.com/news/2020/feb/18/stringers-where-pollution-worst-san-diego/ [last access: 10/28/2022].
Pérez, Laura E.: Spirit glyphs. Reimagining art and artist in the work of Chicana Tlamatinime. In: *Modern Fiction Studies* 44/1 (1998), pp. 36–76.
Piepzna-Samarasinha, Leah Lakshmi: A Not-So-Brief Personal History of the Healing Justice Movement, 2010–2016. In: *MICE Magazine* (2016), https://micemagazine.ca/issue-two/not-so-brief-personal-history-healing-justice-movement-2010-2016 [last access: 01/10/2022].
Pink, Sarah et al.: *Digital Ethnography. Principles and Practice.* London 2016.
Pisarz-Ramirez, Gabriele: *MexAmerica. Genealogien und Analysen postnationaler Diskurse in der kulturellen Produktion von Chicanos/as.* Heidelberg 2005.
Postero, Nancy/Elinoff, Eli: Introduction: A return to politics. In: *Anthropological Theory* 19/1 (2019), pp. 3–28.
Price, David H.: *Weaponizing Anthropology. Social Science in Service of the Militarized State.* Edinburgh 2011.
Pulido, Alberto López/Reyes, Rigoberto "Rigo": *San Diego Lowriders. A History of Cars and Cruising.* Charleston 2017.
Pulido, Laura: Geographies of race and ethnicity III: Settler colonialism and nonnative people of color. In: *Progress in Human Geography* 42/2 (2018), pp. 309–318.
Quijano, Aníbal: Colonialidad del poder y clasificación social. In: *Journal of World-Systems Research* 4/2 (2000), pp. 342–386.
Quijano, Aníbal: Coloniality of Power, Eurocentrism, and Social Classification. In: Moraña, Mabel/Dussel, Enrique/Jáuregui, Carlos A. (Eds.): *Coloniality at Large: Latin America and the Postcolonial Debate.* Durham/London 2008, pp. 181–224.
Raji, Deborah: How our data encodes systematic racism. In: *MIT Technology Review* (10/12/2020), https://www.technologyreview.com/2020/12/10/1013617/racism-data-science-artificial-intelligence-ai-opinion/ [last access: 03/10/2022].
Ramirez, Iskra: Latix, Brujy, Commodification and Community. On: *Masters of Media* (09/24/2018), https://mastersofmedia.hum.uva.nl/blog/2018/09/24/latinx-brujx-commodification-and-community/ [last access: 01/20/2022].

Rancière, Jacques: Politics, Identification, and Subjectivization. In: *The Identity in Question* 61 (1992), pp. 58–64.
Rancière, Jacques: *Das Unvernehmen.* Frankfurt am Main 2002.
Rancière, Jacques: *Dissensus. On Politics and Aesthetics.* London 2010.
Rancière, Jacques: Contemporary Art and the Politics of Aesthetics. In: Hinderliter, B./Kaizen, W./Manssor, J./MacCormack, S. (Eds.): *Communities of Sense: Rethinking Aesthetics and Politics.* Durham/London 2014, pp. 31–50.
Raschig, Megan: Cargas Coming Down: Chronic Stress, Chicana-Indigenous Spiritual Healing, and Feminist Fugitive Potentiality. In: *Feminist Anthropology* (early view). DOI: 10.1002/fea2.12100, 2022.
Ray, Justin: California's central role in the eugenics movement. In: *Los Angeles Times* (07/20/2021), https://www.latimes.com/california/newsletter/2021-07-20/california-eugenics-reparations-sterilization-essential-california?utm_source=headtopics&utm_medium=news&utm_campaign=2021-07-20 [last access: 05/28/2022].
Reeves, Madelaine: Infrastructures of Hope. Anticipating 'Independent Roads' and Territorial Integrity in Southern Kyrgyzstan. In: *Ethnos* 82/4 (2017), pp. 711–737.
Regnier, Denis: Forever slaves? Inequality, uncleanliness and vigilance about origins in the southern highlands of Madagascar. In: *Anthropological Forum* 29/3 (2019), pp. 249–266.
Rendon-Alvarez, Karla: Portside Residents Eligible to Receive Free Air Purifier, Monitor; Here's How. In: *NBC San Diego* (02/10/2020), https://www.nbcsandiego.com/news/local/portside-residents-eligible-to-receive-free-air-purifier-monitor-heres-how/2866461/#:~:text=Those%20who%20would%20like%20to,name%2C%20address%20and%20phone%20number [last access: 10/31/2022].
Repard, Pauline: Driver Who Crashed Off Coronado Bridge, Killing Four, Found Guilty of Manslaughter. On: *The San Diego Union Tribune* (02/13/2019), https://www.sandiegouniontribune.com/news/courts/sd-me-sepolio-coronado-bridge-chicano-verdict-20190209-story.html [last access: 11/14/2022].
Restrepo, Eduardo/Escobar Arturo: 'Other Anthropologies and Anthropology Otherwise'. Steps to a World Anthropologies Framework. In: *Critique of Anthropology* 25/2 (2005), pp. 99–129.
Rivera, Evelya: "Chicanismo." In: Vásquez, Olga A. (ed.): *Chicana/o Art of San Diego.* (Chicana/o Visual Culture Communications 175/198). San Diego 2004, pp. 11 f.
Rodríguez, Marc Simon: *Rethinking the Chicano Movement.* London 2014.
Rosa, Jonathan/Bonilla, Yarimar: Deprovincializing Trump, Decolonizing Diversity, and Unsettling Anthropology. In: *American Ethnologist* 44/2 (2017), pp. 201–208.
Rosa, Jonathan/Díaz, Vanessa: Raciontologies. Rethinking Anthropological Accounts of Institutional Racism and Enactments of White Supremacy in the United States. In: *American Anthropologist* (2019), pp. 1–13.
Rosaldo, Renato: Cultural Citizenship in San José, California. In: *Political and Legal Anthropology Review* 17/2 (1994), pp. 57–63.
Rosen, Martin D./Fisher, James: Chicano Park and the Chicano Park Murals: Barrio Logan, City of San Diego, California. In: *The Public Historian* 23/4 (2001), pp. 91–111.
Rothstein, Richard: *The Colour of Law. A Forgotten History of How Our Government Segregated America.* New York/London 2017.
Said, Edward: *Orientalism.* New York 1978.
Saldívar, José David: Border Thinking, Minoritized Studies, and Realist Interpellations. The Coloniality of Power from Gloria Anzaldúa to Arundhati Roy. In: Alcoff, L.M./Hames-García, M./Mohanty, S.P./Moya, P.M.L (Eds.): *Identity politics reconsidered.* New York 2006, pp. 152–170.

Sánchez, Aaron E.: *Homeland: Ethnic Mexican Belonging since 1900.* Norman 2021.
Sánchez, Rita: Chicanas in the Arts, 1970–1995. With Personal Reflections. In: Griswold del Castillo, Richard (Ed.): *Chicano San Diego. Cultural Space and Struggle for Justice.* Tucson 2007, pp. 158–201.
Sánchez, Rita: Learning from the Past. Some Concluding Comments. In: Griswold del Castillo, Richard (Ed.): *Chicano San Diego. Cultural Space and Struggle for Justice.* Tucson 2007, pp. 250–251.
Sandset, Tony: The necropolitics of COVID-19. Race, class and slow death in an ongoing pandemic. In: *Global Public Health* 16/8–9 (2021), pp. 1411–1423.
Schönwald, Antje: Ein Blick auf Chicanos. Mexikaner und ihre Nachfahren in der amerikanischen Stadt. In: Weber, Florian/Kühne, Olaf (Eds.): *Fraktale Metropolen, Hybride Metropolen.* Wiesbaden 2016, pp. 349–363.
Scott, James W. (Ed.): *A Research Agenda for Border Studies.* Cheltenham 2020.
Scott, James W.: Introduction to a Research Agenda for Border Studies. In: Scott, J.W. (Ed.): *A Research Agenda for Border Studies.* Cheltenham 2020, pp. 3–24.
Shapira, Harel: *Waiting for Jose. The Minutemen's pursuit of America.* Princeton 2013.
Sharpe, Christina: *In the Wake. On Blackness and Being.* Durham 2016.
Silverblatt, Irene Marsha: *Moon, Sun, and Witches: Gender Ideologies and Class in Inca and Colonial Peru.* Princeton 1987.
Simmel, Georg: Bridge and Door: In: *Theory, Culture and Society* 11 (1994 [1909]), pp. 5–10.
Simpson, Audra: *Mohawk Interruptus. Political Life Across the Borders of Settler States.* Durham 2014.
Siry, C./Ali-Khan, C./Zuss, M.: Cultures in the making: An examination of the ethical and methodological implications of collaborative research. In: *Forum: Qualitative Social Research* 12/2 (2011), s.p.
Skinner, Diane: Foucault, Subjectivity and Ethics. Towards a Self-Forming Subject. In: *Organization* 20/6 (2012), pp. 904–923.
Sökefeld, Martin/Strasser, Sabine: Introduction: Under suspicious eyes – surveillance states, security zones and ethnographic fieldwork. In: *Zeitschrift für Ethnologie* 141 (2016), pp. 159–176.
Sotirin, Patty: Introduction to Feminist Vigilance. In: Sotirin, Patty/Bergvall, Victoria L./Shoos, Diane L. (Eds.): *Feminist Vigilance.* ebook 2020.
Sparrow, Glen: San Diego–Tijuana: Not quite a binational city or region. In: *GeoJournal* 54 (2001), pp. 73–83.
Speed, Shannon: At the Crossroads of Human Rights and Anthropology: Toward a Critically Engaged Activist Research. In: *American Anthropologist* 108/1 (2006), pp. 66–76.
Spivak, Gayatri Chakravorty: Can the Subaltern Speak? In: Nelson, Cary/Grossberg, Lawrence (Eds.): *Marxism and the Interpretation of Culture.* Basingstoke 1988, pp. 271–313.
Star, Susan Leigh: The Ethnography of Infrastructure. In: *American Behavioral Scientist* 43/3 (1999), pp. 377–391.
Statista: Ranking der 15 Länder mit den meisten Migranten weltweit im Jahr 2020. In: *Statista.com* (12/21/2021), https://de.statista.com/statistik/daten/studie/185928/umfrage/groesste-einwanderungslaender/ [last access: 01/23/2022].
St John, Rachel: *Line in the Sand: A History of the Western U.S.-Mexico Border.* Princeton/New Jersey 2011.
Stop LAPD Spying!: Automating Banishment: The Surveillance and Policing of Looted Land. On: *Stop LAPD Spying!* (11/2021), https://automatingbanishment.org/ [last access: 03/10/2022].

Sunjata, John Kamaal: Gentrification as Settler-colonialism: Urban Resistance against Urban Colonization. In: *Hampton* (11/09/2021), https://www.hamptonthink.org/read/gentrification-as-settler-colonialism-urban-resistance-against-urban-colonization [last access: 10/31/2022].

Sze, Julie: *Environmental Justice in a Moment of Danger.* Berkeley, CA 2020.

Takvorian, Diane: Opinion: Plans to Reduce Pollution Aren't Enough for Barrio Logan. Local Government Must Do More. On: *The San Diego Union Tribune* (02/03/2022), https://www.sandiegouniontribune.com/opinion/commentary/story/2022-02-03/barrio-logan-diesel-pollution-clean-air?utm_source=dlvr.it&utm_medium=twitter [last access: 10/28/2022].

Talamantez, Josephine: *Chicano Park and the Chicano Park Murals: A National Register Nomination.* (MA thesis, History). Sacramento 2011.

Talante, Julia: Fungible Indigeneity and Blackness: On the Paradigmatic Domination of Borderlands Theory. In: *Medium* (06/11/2018), https://juliatalante.medium.com/fungible-indigeneity-and-blackness-on-the-paradigmatic-domination-of-borderlands-theory-c4084fbfe919 [last access: 08/31/2022].

Taylor, Dorceta E.: *Toxic Communities: Environmental Racism, Industrial Pollution, and Residential Mobility.* New York 2014.

The San Diego Union-Tribune Editorial Board: Will new air pollution study finally lead to relief for Barrio Logan? On: *The San Diego Union-Tribune* (01/30/2019), https://www.sandiegouniontribune.com/opinion/editorials/sd-barrio-logan-20190129-story.html [last access: 10/28/2022].

The 2021 Barrio Logan community plan. In: *Plan Barrio* (2022), https://www.planbarrio.org/ [last access: 10/31/2022].

Thin, Neil: On the Primary Importance of Secondary Research. In: *Doing Anthropological Research: A Practical Guide* X (2014), pp. 37–54.

Trottier, Daniel: Digital Vigilantism as Weaponisation of Visibility. In: *Philosophy & Technology* 30 (2017), pp. 55–72.

Trottier, Daniel: Denunciation and doxing: towards a conceptual model of digital vigilantism. In: *Global Crime* 21/3–4 (2020), pp. 196–212.

Tuck, Eve/Yang, K. Wayne: Decolonization is not a metaphor. In: *Decolonization: Indigeneity, Education & Society* 1/1 (2012), pp. 1–40.

Turunen, J./Čeginskas, V.L.A./Kaasik-Krogerus, S./Lähdesmäki, T./Mäkine, K.: Poly-Space: Creating new concepts through reflexive team ethnography. In: *Challenges and Solutions in Ethnographic Research. Ethnography with a Twist* (2021), pp. 3–20.

Valencia Triana, Sayak: *Capitalismo Gore. Control económico, violencia y narcopoder.* Barcelona 2010.

Valenzuela-Levi, Nicolas: "Somos Zona Roja": institutionalised exclusion from broadband internet services in Santiago de Chile. In: Alderman, J./Goodwin, G. (Eds.): *The Social and Political life of Latin American Infrastructure. Meanings, Values, and Competing Visions of the Future.* London 2022.

Van Houtum, Henk/Van Naerssen, Ton: Bordering, ordering and othering. In: *Tijdschrift voor economische en sociale geografie* 93/2 (2002), pp. 125–136.

Vargas, Zaragosa: *Crucible of Struggle. A History of Mexican Americans from Colonial Times to the Present Era.* Oxford 2017.

Vasconcelos, José: *La Raza Cósmica.* México 2010.

Venkatraman, Sakshi: Anti-Asian hate crimes rose 73% last year, updated FBI data says. On: NBC News (10/25/2021), https://www.nbcnews.com/news/asian-america/anti-asian-hate-crimes-rose-73-last-year-updated-fbi-data-says-rcna3741 [last access: 10/28/2022].

Veracini, Lorenzo: *Settler Colonialism. A Theoretical Overview.* New York 2010.

Verdery, Katherine: Observers Observed. In: *Anthropology Now* 4/2 (2012), pp. 14–23.
Vigh, Henrik: Vigilance. On Conflict, Social Invisibility, and Negative Potentiality. In: *Social Analysis* 55/3 (2011), pp. 93–114.
Vila, Pablo: The Polysemy of the Label "Mexican" on the Border. In: *Ethnography at the Border* X (2003), pp. 105–140.
Vizenor, Gerald: *Fugitive poses: Native American Indian scenes of absence and presence.* Lincoln 1998.
Vizenor, Gerald: *Manifest Manners: Narratives on Postindian Survivance.* Lincoln 1999.
Vizenor, Gerald/Lee, A. Robert: *Postindian Conversations.* Lincoln 1999.
von Schnitzler, Antina: Infrastructure, Apartheid Technopolitics, and Temporalities of "Transition." In: Appel, Hannah/Anand, Nikhil/Gupta, Akhil (Eds.): *The Promise of Infrastructure.* Durham/London 2018.
Wallis, George W.: Chronopolitics. The Impact of Time Perspectives on the Dynamics of Change. In: *Social Forces* 49/1 (1970), pp. 102–108.
Walsh, Catherine: "Other" Knowledges, "Other" Critique. Reflections on the Politics and Practices of Philosophy and Decoloniality in the "Other America." In: *Transmodernity: Journal of Peripheral Cultural Production of the Luso-Hispanic World* 1/1 (2012), pp. 11–27.
Walsh, James: Community, surveillance and border control. The case of the minuteman project. In: Deflem, M./Ulmer, J.T. (Eds.): *Surveillance and Governance. Crime Control and Beyond.* Bingley 2008, pp. 9–34.
Watts, Brenda: Aztlán as a Palimpsest. From Chicano Nationalism toward Transnational Feminism in Anzaldúa's Borderlands. In: *Latino Studies* 2/3 (2004), pp. 304–321.
Whittaker, Catherine: Felt power. Can Mexican Indigenous women finally be powerful? In: *Feminist Anthropology* 1/2 (2020), pp. 288–303.
Whittaker, Catherine/Dürr, Eveline: Vigilance, Knowledge, and De/Colonization. Protesting While Latin@ in the US-Mexico Borderlands. In: *Conflict and Society: Advances in Research* 8/1 (2022), pp. 156–171. DOI: 10.3167/arcs.2022.080110.
Wilk, Richard: Colonial Time and TV Time. Television and Temporality in Belize. In: *Visual Anthropology Review* 10 (1994), pp. 94–102.
Williams, David R./Mohammed, Selina A.: Racism and Health I: Pathways and Scientific Evidence. In: *American Behavioral Scientist* 57/8 (2013), pp. 1152–1173.
Wolf-Meyer, Matthew: Editorial Introduction. Alertness, or the Other Side of Sleep. In: *Anthropology of Consciousness* 24/2 (2013), pp. 93–95.
Wolverton, Joe, II: "Sousveillance": When the Watched Become the Watchers. In: *The New American* (11/07/2012), https://thenewamerican.com/sousveillance-when-the-watched-become-the-watchers/ [last access: 10/31/2022].
Wood Bartholomew, Melissa/Harris, Abril N./Maglalang, Dale Dagar: A Call to Healing: Black Lives Matter Movement as a Framework for Addressing the Health and Wellness of Black Women. In: *Community Psychology in Global Perspective* 4/2 (2018), pp. 85–100.
Wrigley-Field, Elizabeth: US racial inequality may be as deadly as COVID-19. In: *PNAS* 117/36 (2020), pp. 21854–21856.
Yam, Kimmy: Jose Antonio Vargas on how aspects of the pandemic mirror struggles of undocumented people. On: *NBC News* (05/01/2020), https://www.nbcnews.com/news/asian-america/jose-antonio-vargas-how-aspects-pandemic-mirror-struggles-undocumented-people-n1197616 [last access: 10/31/2022].
Yeh, Rihan: *Passing. Two publics in a Mexican border city.* Chicago 2018.
Zamora, Beatriz: *The Spirit of Chicano Park / El Espiritú del Parque Chicano.* San Diego 2020.

Zaragocin, Sofia: Decolonial Feminist Geography. In: *Geopauta* 4/4 (2020), pp. 18–30.
Zevely, Jeff: Why a drone may save your life sometime soon in San Diego. On: *CBS8* (01/30/2020), https://www.cbs8.com/article/news/local/zevely-zone/san-diego-police-department-uses-tactical-drone-solutions/509-83f3d56f-2cb6-40ab-b8ec-2b2e41390539 [last access: 08/15/2022].
Zukin, Sharon: *Naked city. The death and life of authentic urban places.* Oxford/New York 2010.

List of Acronyms

ICE	Immigration and Customs Enforcement
FBI	Federal Bureau of Investigation
IRCA	Immigration Reform and Control Act
INS	Immigration and Naturalization Service
NAFTA	North American Free Trade Agreement
CIA	Central Intelligence Agency
CPSC	Chicano Park Steering Committee
UCSD	University of California San Diego
SDSU	San Diego State University
LMU	Ludwig Maximilian University (Munich)
KKK	Ku Klux Klan
NASSCO	National Steel and Shipbuilding Company
NSSF	National Shooting Sports Foundation
EHC	Environmental Health Coalition
SDGND	San Diego Green New Deal

List of Figures and Illustrations

List of Figures

Preface
Figure 1: From left to right, Catherine Whittaker, Jonathan Alderman (back), Eveline Dürr (front), Carolin Luiprecht (photo credit: Eveline Dürr).

Chapter 1
Figure 2: Public Service Announcement on a San Diegan trolley, February 2020 (photo credit: Catherine Whittaker).
Figure 3: Barrio Logan gateway sign (photo credit: Jonathan Alderman).
Figure 4: Barrio Logan trolley station mural: "El pueblo unido jamás será vencido (A united people will never be defeated.)" (photo credit: Catherine Whittaker).

Chapter 3
Figure 5: The Coronado Bridge (photo credit: Jonathan Alderman).
Figure 6: The mural of the Coronado Bridge in the Barrio Dogg diner (photo credit: Catherine Whittaker).
Figure 7: Mural commemorating the creation of Chicano Park and memorial to the people killed in Chicano Park (photo credit: Catherine Whittaker).
Figure 8: The mural of Coatlicue (photo credit: Catherine Whittaker).

Chapter 4
Figure 9: "Somos un Pueblo sin Fronteras" banner in the Centro Cultural de la Raza (photo credit: Catherine Whittaker).
Figure 10: Map of the United States showing the area pertaining to Aztlán (photo credit: Catherine Whittaker).

Chapter 5
Figure 11: *WHY US*, mural, Chicano Park (photo credit: Catherine Whittaker).
Figure 12: Flow chart hanging outside the *brujxs'* shop (photo credit: Catherine Whittaker).

Conclusion
Figure 13: A tote bag sold on Logan Avenue, Barrio Logan, in August 2022 (photo credit: Catherine Whittaker).

List of Illustrations

Chapter 1-7 and front cover
Illustrations: Nanzi Muro, *Untitled*, 2022

List of Maps

Chapter 1
Map 1: Red represents white, blue represents Black, green represents Asian, orange represents Hispanic, yellow represents other. Each dot is 25 people. Data is from Census 2010. https://commons.wikimedia.org/wiki/File:Race_and_ethnicity_2010-_San_Diego_ (5560483270).png, link to the license: https://creativecommons.org/licenses/by-sa/2.0/deed.en.
Map 2: Map of the main research locations in the city of San Diego (credit: Daniel Dumas).

Chapter 5
Map 3: Toxic emissions in San Diego according to neighborhood (source: OEHHA, Toxic Releases from Facilities). OEHHA California Office of Environmental Health Hazard Assessment: "Toxic Releases from Facilities." n.d. [last access: 08/30/2022].
Map 4: Disadvantaged communities in San Diego (source: OEHHA, SB 535 Disadvantaged Communities). OEHHA California Office of Environmental Health Hazard Assessment: "Sb 535 Disadvantaged Communities." n.d. [last access: 08/30/2022].

Lists of Charts

Chapter 1
Chart 1: Five largest ethnic groups in San Diego according to census 2020. Created by Carolin Luiprecht, data from https://datausa.io/profile/geo/san-diego-ca (retrieved on 26/08/2022).
Chart 2: Hispanic population in San Diego according to census 2020. Created by Carolin Luiprecht, data from https://datausa.io/profile/geo/san-diego-ca (retrieved on 26/08/2022).

www.ingramcontent.com/pod-product-compliance
Lightning Source LLC
Chambersburg PA
CBHW050535300426
44113CB00012B/2109